BASIC
BLUEPRINT READING
FOR
PRACTICAL APPLICATIONS

BASIC
BLUEPRINT READING
FOR
PRACTICAL APPLICATIONS

BY JOHN E. TRAISTER

TAB BOOKS Inc.
BLUE RIDGE SUMMIT, PA. 17214

FIRST EDITION
SECOND PRINTING

Printed in the United States of America

Library of Congress Cataloging in Publication Data

Traister, John E.
Basic blueprint reading for practical applications.

Includes index.
1. Blue-prints.   I. Title.
T379.T7   1983   604.2′5   82-19329
ISBN 0-8306-0246-1
ISBN 0-8306-0146-5 (pbk.)

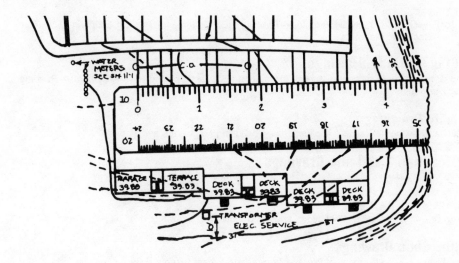

Contents

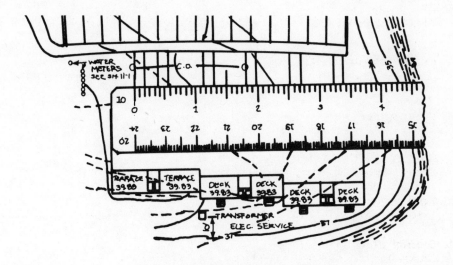

Introduction

INTEREST IN DO-IT YOURSELF PROJECTS IS AT ITS highest. People have more leisure time. The spiraling cost of labor and materials has brought with it the need for more do-it-yourselfers, all of whom should know how to interpret working drawings and specifications of various projects. These include blueprints for homes and buildings as well as plans for smaller projects such as a workbench, a garden trellis, and the like. Those involved in constructing electronic projects, for example, must know the meaning of each symbol and the relationship of each one to the entire project. Almost everyone has to interpret some type of drawing daily. Nearly every object purchased these days is only partially assembled when it comes out of the box. Drawings and instructions are included showing how the object is to be assembled. The owner must be able to "read" them.

The material presented in this book is primarily designed to acquaint homeowners and do-it-yourselfers with blueprint reading as applied to practical applications they will more than likely encounter. This book is also a good text for students, workers, and technicians entering the construction or technology fields.

Chapter 1

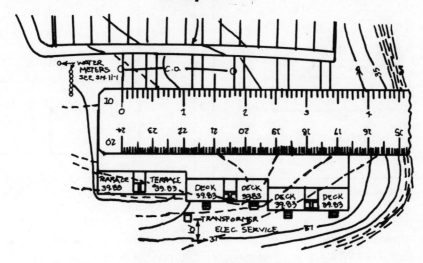

Overview

I F YOU ARE INVOLVED IN BUILDING OR ASSEM-
bling a project, you will encounter many
types of drawings. A brief sampling of each
follows; each type is covered in more detail
in later chapters along with practical applications.

PICTORIAL DRAWINGS

In these drawings an object is drawn in one
view only. Three-dimensional effects are
simulated on the flat plane of drawing paper
by drawing several faces of an object in a
single view. These drawings are used to
convey information to those people who are
not trained in blueprint reading or to supplement the conventional orthographic drawings in more complex systems.

The pictorial drawings most often found
in plans for home projects include:

● Isometric drawing
● Oblique drawing
● Perspective drawing

The oblique drawing will be found most
often. An oblique drawing of a workbench (in
an exploded view) is shown in Fig. 1-1. This
drawing clearly shows how the bench is
constructed and approximately what it will
look like once it is completed. The main disadvantage of pictorial drawings is that intricate parts cannot be pictured clearly and are
difficult to dimension.

ORTHOGRAPHIC PROJECTION DRAWINGS

These drawings are used more than any
other in plans for homes and smaller projects. *Orthographic projection* drawings
generally give all plan views, elevation
views, dimensions, and other details necessary to construct a project or object.

To illustrate the practicality of the or-

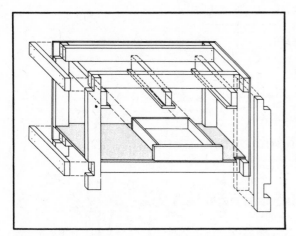

Fig. 1-1. An exploded view of a workbench using the oblique technique.

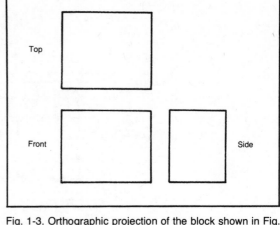

Fig. 1-3. Orthographic projection of the block shown in Fig. 1-2.

thographic drawing, look at the pictorial drawing in Fig. 1-2. While this view clearly suggests the form of a block, it does not show the actual shape of the surfaces. Also, it does not show the dimensions of the object so that it may be constructed.

An orthographic projection of this same

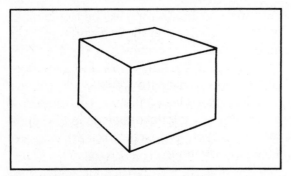

Fig. 1-2. Pictorial drawing of a cube or block.

block appears in Fig. 1-3. The front view in Fig. 1-3 shows the block as though you were looking straight at the front. Another view is shown as though you were looking straight at the left side, right side, rear, and top. These views, when combined with dimensions, will allow you to construct the object properly from metal, wood, plastic, etc. The

material may be specified in written specifications, a schedule, or by merely adding a note to the drawing.

DIAGRAMS

Diagrams are drawings that are intended to show components and their related connections in diagrammatic form. Such drawings are seldom drawn to scale and show only the working association of the different components. Symbols are used extensively in diagrams to represent the various pieces of equipment or components. Lines are used to connect these symbols—indicating the size, type, and number of components or pieces of equipment. Diagrams are used extensively in electronic and electrical projects (Fig. 1-4) and plumbing projects.

Construction working drawings consist of lines, symbols, dimensions, and notations to accurately convey an idea or design so that the project or object may be built. A working drawing is an abbreviated language for conveying a large amount of exact, detailed information, which would otherwise take many pages of manuscript or hours of verbal instruction to convey.

2

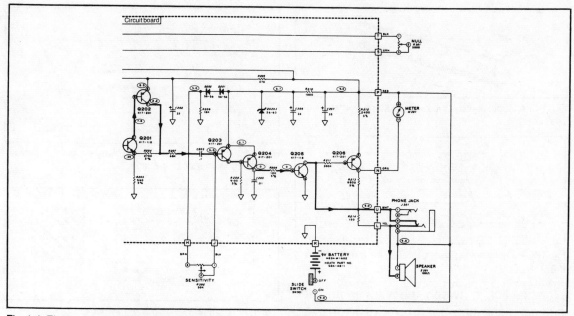

Fig. 1-4. Electronic schematic diagram used for constructing a project and also for troubleshooting any problems.

SECTIONAL VIEWS

A *section* of any object—as applied to construction drawings—is what could be seen if the object was sliced or sawed into two parts at the point where the section was taken. If you wanted to see how a golf ball is constructed, you could place the ball in a vise and cut it in half with a hacksaw. When the two parts are separated, you can easily see how the ball is constructed, or you would at least have a view of its internal construction.

You must use a considerable amount of visualization in dealing with sections. There are no given rules for determining what a section will look like. If a piece of galvanized water pipe is cut vertically, the section will appear as shown in Fig. 1-5. If cut horizon-

Fig. 1-6. A section cut horizontally through a water pipe will appear as a circle on the drawing.

tally, the section will appear as shown in Fig. 1-6. If cut on a slant, the section will form an ellipse as shown in Fig. 1-7.

The theory of construction for a sectional view is illustrated in Figs. 1-8 through 1-10. Figure 1-8 shows a pipe view from the end and side. In Fig. 1-9 a cutting plane is shown passing through the pipe; that is where the section is taken and is indicated by the cutting plane line. The portion of the

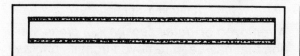

Fig. 1-5. A vertical section through a piece of water pipe will appear as a rectangle on a drawing.

Fig. 1-7. A section through a water pipe cut on a slant will form an ellipse.

3

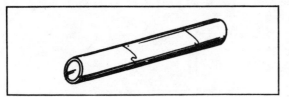

Fig. 1-8. A length of water pipe viewed at the end and side.

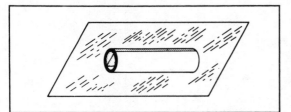

Fig. 1-9. A cutting plane passing through the length of pipe in Fig. 1-8.

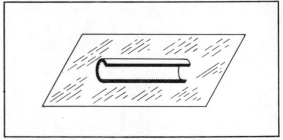

Fig. 1-10. The closest section of pipe has been removed, revealing the interior of the pipe.

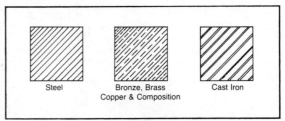

Fig. 1-11. Common section lining symbols.

pipe section between the viewer and the cutting plane has been removed to reveal the interior details of the pipe. In Fig. 1-10 the cutting plane is removed, and the section would be drawn in an orthographic view.

Section lining or cross-hatching is used in sectional views to indicate the various materials of construction. For common metal such as the galvanized steel of the pipe, lining is made with fine lines, usually drawn at angles of 45 degrees. See Fig. 1-11.

Other sectional views are classified as:

- Full section.
- Half section.
- Revolved section.
- Removed section.
- Aligned section
- Broken-out section.
- Assembly section.

Full Section

A *full section* is a view in which the cutting plane is assumed to pass entirely through the object. The sectional view in Fig. 1-12 is a full section.

Half Section

A *half section* is a sectional view in which the cutting plane passes halfway through the object. One half of the view is shown in section, while the other half is shown from the exterior. Figure 1-13 shows a cutting plane passing halfway through a piece of under-floor round duct that is constructed of tile. Figure 1-14 shows a portion in front of the cutting plane removed.

Revolved Section

A *revolved section* is a cross section that has been revolved through 90 degrees. It is used to show the true shape of the cross section of bars and other elongated parts. Figure 1-15 shows such a section.

Removed Section

A *removed* or *detail section* is a cross section that has been removed from its original position to a convenient space near the principal view. See Fig. 1-16.

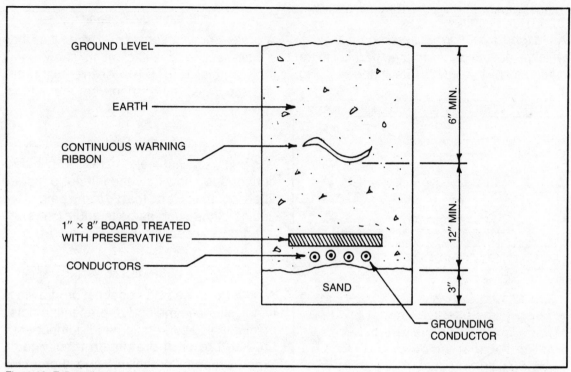

GROUND LEVEL

EARTH

CONTINUOUS WARNING RIBBON

1″ × 8″ BOARD TREATED WITH PRESERVATIVE

CONDUCTORS

SAND

6″ MIN.

12″ MIN.

3″

GROUNDING CONDUCTOR

Fig. 1-12. Full section.

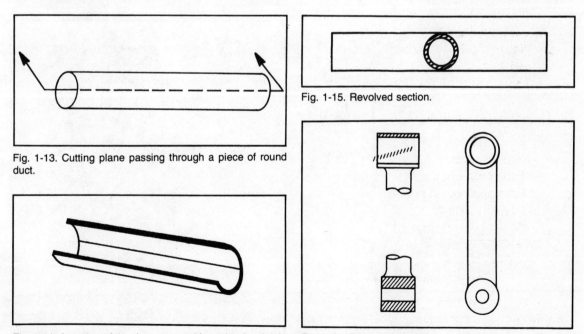

Fig. 1-13. Cutting plane passing through a piece of round duct.

Fig. 1-14. A portion of the duct removed from the drawing in Fig. 1-13.

Fig. 1-15. Revolved section.

Fig. 1-16. Removed section.

Aligned Section

An *aligned section* is a sectional view in which a sloping part is rotated parallel to the cutting plane to show its true shape. See Fig. 1-17.

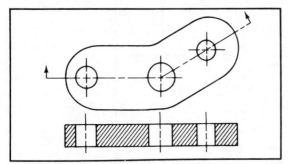

Fig. 1-17. Aligned section.

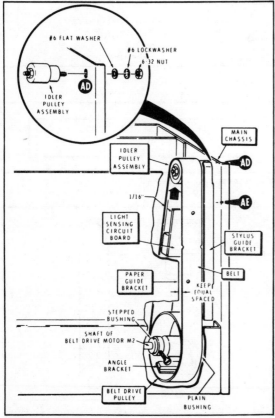

Fig. 1-18. Broken-out section (courtesy Heath Company, Benton Harbor, MI 49022).

Broken-Out Section

A *broken-out section* is used when less than a half section is sufficient to show some interior detail (Fig. 1-18). An irregular break line separates the section from the exterior view.

Assembly Section

Assembly drawings often are drawn in section form to show how the interior parts are fitted together. Parts that lie in the path of the cutting plane such as bolts and screws are not drawn in section. See Fig. 1-19.

SCHEDULES

A *schedule* used on construction drawings is a systematic method of presenting notes or materials lists, equipment, components, and the like on a drawing in tabular form. When properly organized and thoroughly understood, schedules are great time-

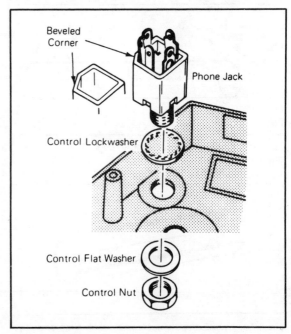

Fig. 1-19. Assembly section.

KEY No.	HEATH Part No.	QTY.	DESCRIPTION	CIRCUIT Comp. No.
ELECTRICAL COMPONENTS				
A1	25-857	1	1500 μF electrolytic capacitor	C1
B1	54-969	1	Power transformer	T1
B2	58-19	1	Yoke	T2
B3	60-54	1	120/240 slide switch	SW1
B4	60-608	1	NOR/LCW slide switch	SW2
B5	60-619	1	Rocker switch	SW3
C1	423-11	1	Fuseholder	
C2	401-163	1	Speaker	SP1
C3	420-607	1	Fan	A1
C4	51-200	1	Flyback transformer	T202
C5	51-197	1	Driver transformer	T201
C6	11-53	1	500 Ω control	R216
HARNESS — CABLES — WIRE — SLEEVING				
	89-54	1	Line cord	
D1	134-1066	1	Flat cable	
D2	134-1067	1	Harness	
D3	134-1075	1	Interconnect cable	
	340-8	48"	Bare wire	

KEY No.	HEATH Part No.	QTY.	DESCRIPTION	CIRCUIT Comp. No.
Harness — Cables — Wire — Sleeving (cont'd.)				
	343-11	8 ft	Large shielded cable	
	343-15	3 ft	Small shielded cable	
	344-15	10 ft	Black stranded wire	
	344-33	36"	Black solid wire	
	344-16	43"	Red stranded wire (thin insulation)	
	344-183	42"	Red stranded wire (thick insulation)	
	344-80	45"	Orange stranded wire	
	344-81	7"	Violet stranded wire	
	344-82	7"	White stranded wire	
	344-126	48"	Brown stranded wire	
	344-154	7"	Yellow stranded wire	
	344-155	73"	Green stranded wire	
	344-156	7"	Blue stranded wire	
	346-77	1"	Large Teflon* sleeving (white)	
	346-64	3"	Large sleeving	
	346-35	6"	Black heat shrinkable sleeving	
	346-21	1"	Small Teflon sleeving (white)	
	346-76	6"	Clear sleeving	
	347-55	24"	8-wire ribbon cable	
	347-60	8 ft	2-wire gray cable	

*Registered Trademark, DuPont

saving items for those preparing the drawings. Schedules also help those people using the drawings to save much valuable time.

Table 1-1 lists the approximate quantities of materials needed to complete an electronic kit project. Similar schedules are provided on woodworking projects, construction projects, etc.

DRAWING SYMBOLS

A working drawing will more than likely be used to show how a certain object, piece of equipment, or system is to be constructed, installed, modified, or repaired. In preparing these drawings symbols are used to simplify the work of the draftsmen and also to keep the size and bulk of the drawings to a workable minimum. Can you imagine the time needed to prepare a set of working drawings for an electronic project if every component had to be drawn as it would be seen with the eye? Consider also the size and number of pages that would be required to hold all these detailed drawings. Symbols used on construction drawings may be considered time-saving devices that convey detailed information from the designer to the person building the project.

The three most common symbols used to indicate earth, rock, and stone fill are shown in Fig. 1-20. An example of their

Fig. 1-20. The three most common symbols used to indicate earth, rock, and stone fill.

practical use is shown in Fig. 1-21. Notice that the entire earth area in Fig. 1-21 is not marked with the earth symbol. Only the areas around other materials are indicated by the symbol.

Symbols for concrete and related materials such as concrete block, terrazzo, etc., are shown in Fig. 1-22. Examples of their use appear in Fig. 1-23.

Several symbols are used to indicate metal. The type employed is usually governed by the size of the drawing or to the scale to which the drawing is done. When the scale used is too small to indicate the type of metal by symbols, notes are normally provided near the items to indicate the type

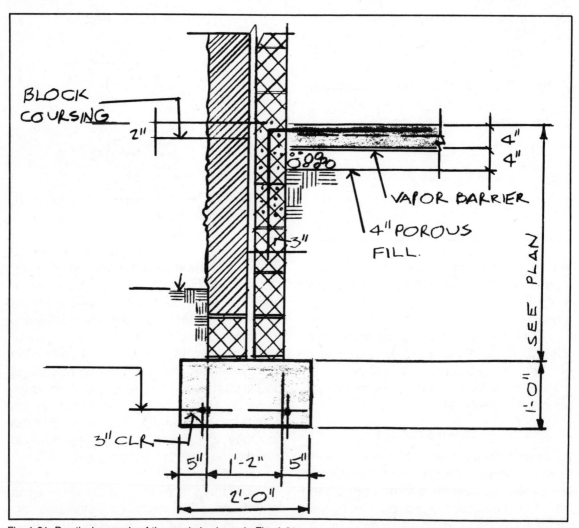

Fig. 1-21. Practical example of the symbols shown in Fig. 1-20.

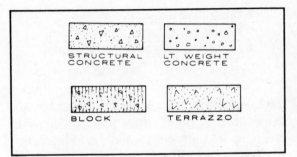

Fig. 1-22. Symbols for concrete and related materials.

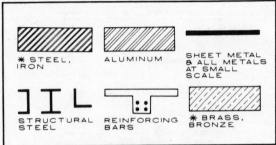

Fig. 1-24. A few of the various symbols used to indicate metal.

of metal to use. A few symbols for metals are shown in Fig. 1-24.

Wood symbols appear in Fig. 1-25. The size of the drawing determines to a great

extent the type of symbol used.

Symbols commonly used to indicate stone and brick are shown in Fig. 1-26. Practical applications of some of the sym-

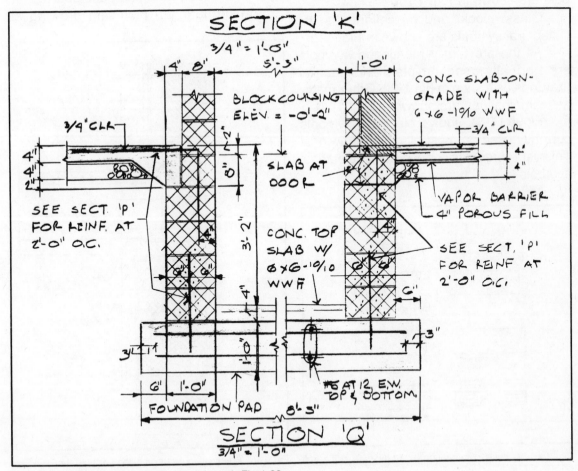

Fig. 1-23. Practical example of the symbol shown in Fig. 1-22.

9

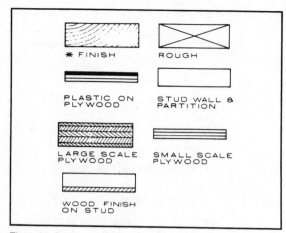

Fig. 1-25. Some symbols used on drawings to indicate wood.

bols are shown in Fig. 1-27.

Glass, block, and miscellaneous architectural symbols are shown in Fig. 1-28.

All the previously mentioned symbols are normally used on plan and section drawings as shown in Fig 1-29. Other material combinations are shown in Fig. 1-30.

Figure 1-31 shows electrical symbols currently used on most electrical drawings. This list represents a good set of electrical symbols in that they are:

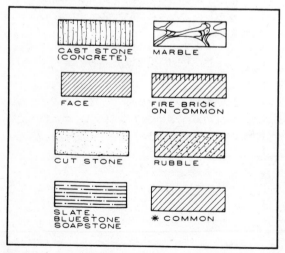

Fig. 1-26. Symbols commonly used to indicate stone and brick.

● Easy to draw.
● Easily interpreted by workmen.
● Sufficient for most applications.

Many symbols have the same basic form, but their meanings differ slightly because of the addition of a line, mark, or abbreviation. A good procedure to follow in learning the different symbols is to first learn the basic form, then apply the variations of that form to obtain the different meanings.

Some symbols contain abbreviations such as "WT" for watertight and "S" for switch. Others are simplified pictographs such as the symbol for safety switch or the symbol for flush-mounted panelboard.

In preparing a plumbing drawing all

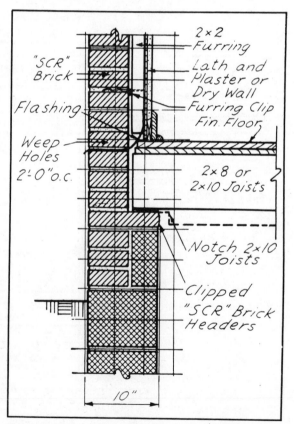

Fig. 1-27. Practical applications of the symbols shown in Fig. 1-26.

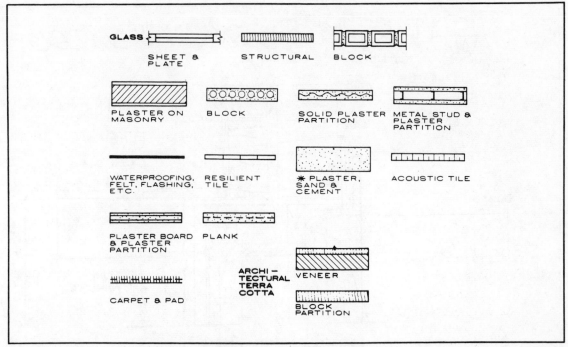

GLASS
SHEET & PLATE

STRUCTURAL

BLOCK

PLASTER ON MASONRY

BLOCK

SOLID PLASTER PARTITION

METAL STUD & PLASTER PARTITION

WATERPROOFING, FELT, FLASHING, ETC.

RESILIENT TILE

* PLASTER, SAND & CEMENT

ACOUSTIC TILE

PLASTER BOARD & PLASTER PARTITION

PLANK

CARPET & PAD

ARCHI- TECTURAL TERRA COTTA

VENEER

BLOCK PARTITION

Fig. 1-28. Symbols used to indicate various architectural materials.

pipe, fittings, fixtures, valves, and other components are shown by symbols such as those in Fig. 1-32. The use of these symbols simplifies considerably the preparation of piping drawings and conserves much time and effort on the part of the installer.

The main purpose of HVAC (heating, ventilating, and air conditioning) drawings is to show the location of the heating, cooling, and air conditioning units along with the related ductwork and piping. Graphical symbols used on HVAC drawings are similar to those used for plumbing, with a few exceptions. See Fig. 1-33.

Notice that the ductwork shown in Fig. 1-34 uses two types of symbols. The main duct consists of a rectangle to represent the actual size of the ductwork (drawn to scale) including reductions in the size of the duct. The branches are shown by a single line with the dimensions of each noted adjacent to

these lines. You will find both methods used on HVAC drawings for house plans—either singly or in combination. Obviously, the two-line method is easier to use, because the ductwork appears as it would if the viewer were looking down on the system from above.

DRAWING DIMENSIONS

A drawing is expected to convey exact information regarding every detail so that an object can be fabricated. It would be impossible to achieve successful results without definite dimensions on a drawing.

A drawing dimension is a numerical value expressed in appropriate units like feet and inches, metric units, etc. They are indicated on drawings in conjunction with lines, symbols, and notes to define the geometrical characteristics of an object.

Dimensions are usually shown between

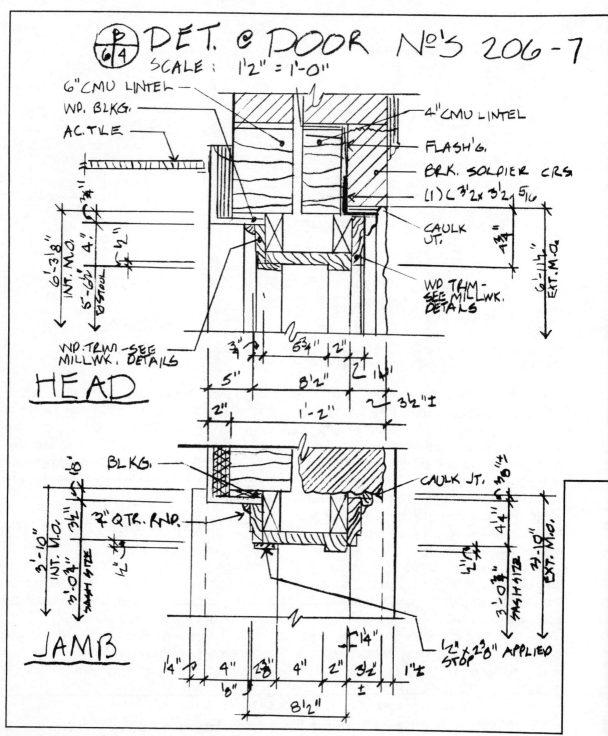

Fig. 1-29. Practical applications of the symbols shown in Fig. 1-28.

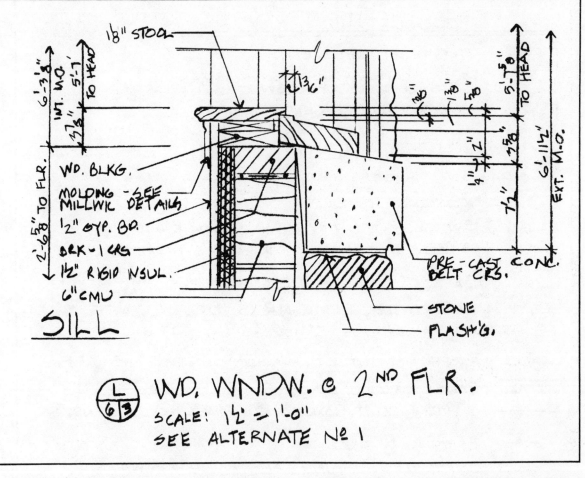

1⅛" STOOL

1 3/16"

6'-3" WT. M.O.
5'-1" TO HEAD
1⅜"

5'-5" TO HEAD
1"
2"
3⅜"
6'-11½" EXT. M.O.
2"
4"
7½"

2'-68" 2'-5" TO FLR.

WD. BLKG.
MOLDING - SEE MILLWK DETAILS
½" GYP. BD.
BRK - 1 CRS
1½" RIGID INSUL.
6" CMU

PRE-CAST CONC. BELT CRS.

STONE
FLASH'G.

SILL

Ⓛ 6/3 WD. WNDW. @ 2ND FLR.
SCALE: 1½" = 1'-0"
SEE ALTERNATE № 1

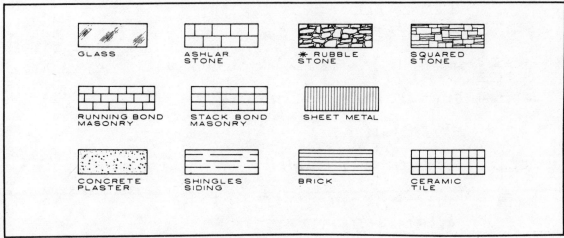

GLASS

ASHLAR STONE

✳ RUBBLE STONE

SQUARED STONE

RUNNING BOND MASONRY

STACK BOND MASONRY

SHEET METAL

CONCRETE PLASTER

SHINGLES SIDING

BRICK

CERAMIC TILE

Fig. 1-30. Material combinations used on working drawings.

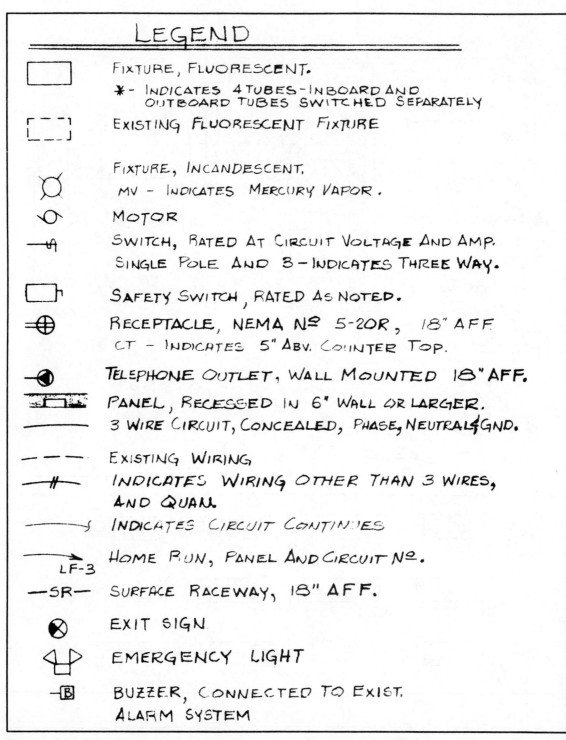

LEGEND

☐ FIXTURE, FLUORESCENT.
  ✳ - INDICATES 4 TUBES-INBOARD AND OUTBOARD TUBES SWITCHED SEPARATELY

┌┄┄┐ EXISTING FLUORESCENT FIXTURE

⊘ FIXTURE, INCANDESCENT.
  MV - INDICATES MERCURY VAPOR.

⊙ MOTOR

—ⓐ SWITCH, RATED AT CIRCUIT VOLTAGE AND AMP.
  SINGLE POLE AND 3 - INDICATES THREE WAY.

☐ SAFETY SWITCH, RATED AS NOTED.

⊕ RECEPTACLE, NEMA № 5-20R, 18" AFF.
  CT - INDICATES 5" ABV. COUNTER TOP.

◖ TELEPHONE OUTLET, WALL MOUNTED 18" AFF.

▬ PANEL, RECESSED IN 6" WALL OR LARGER.
— 3 WIRE CIRCUIT, CONCEALED, PHASE, NEUTRAL & GND.

- - - EXISTING WIRING

—╫— INDICATES WIRING OTHER THAN 3 WIRES, AND QUAN.

——√ INDICATES CIRCUIT CONTINUES

——→ HOME RUN, PANEL AND CIRCUIT №.
LF-3

—SR— SURFACE RACEWAY, 18" AFF.

⊗ EXIT SIGN.

⬡ EMERGENCY LIGHT

—Ⓑ BUZZER, CONNECTED TO EXIST. ALARM SYSTEM

Fig. 1-31. A list of electrical symbols currently in use.

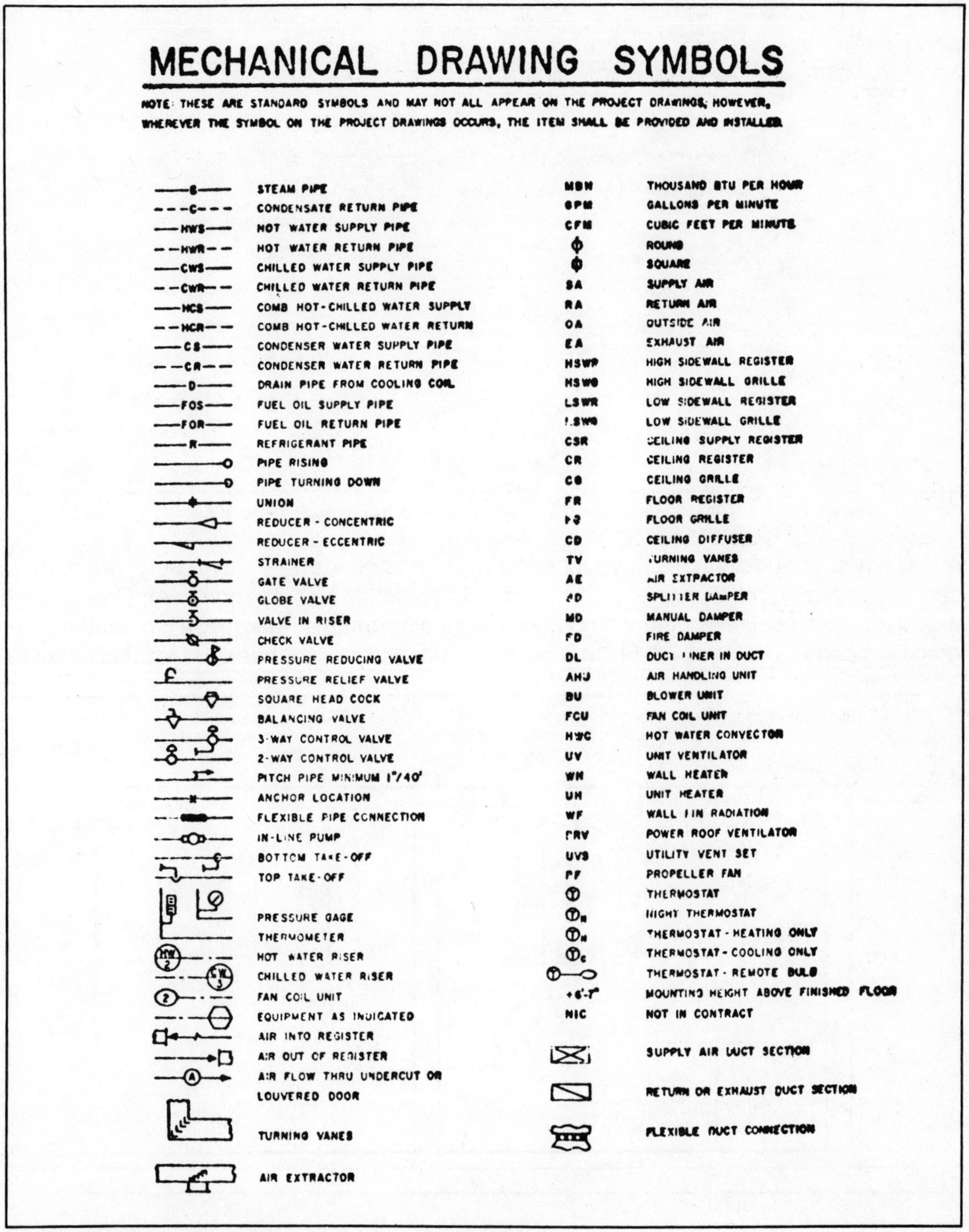

MECHANICAL DRAWING SYMBOLS

NOTE: THESE ARE STANDARD SYMBOLS AND MAY NOT ALL APPEAR ON THE PROJECT DRAWINGS; HOWEVER, WHEREVER THE SYMBOL ON THE PROJECT DRAWINGS OCCURS, THE ITEM SHALL BE PROVIDED AND INSTALLED.

Symbol	Description	Symbol	Description
S	STEAM PIPE	MBH	THOUSAND BTU PER HOUR
C	CONDENSATE RETURN PIPE	GPM	GALLONS PER MINUTE
HWS	HOT WATER SUPPLY PIPE	CFM	CUBIC FEET PER MINUTE
HWR	HOT WATER RETURN PIPE	⌀	ROUND
CWS	CHILLED WATER SUPPLY PIPE	□	SQUARE
CWR	CHILLED WATER RETURN PIPE	SA	SUPPLY AIR
HCS	COMB HOT-CHILLED WATER SUPPLY	RA	RETURN AIR
HCR	COMB HOT-CHILLED WATER RETURN	OA	OUTSIDE AIR
CS	CONDENSER WATER SUPPLY PIPE	EA	EXHAUST AIR
CR	CONDENSER WATER RETURN PIPE	HSWR	HIGH SIDEWALL REGISTER
D	DRAIN PIPE FROM COOLING COIL	HSWG	HIGH SIDEWALL GRILLE
FOS	FUEL OIL SUPPLY PIPE	LSWR	LOW SIDEWALL REGISTER
FOR	FUEL OIL RETURN PIPE	LSWG	LOW SIDEWALL GRILLE
R	REFRIGERANT PIPE	CSR	CEILING SUPPLY REGISTER
	PIPE RISING	CR	CEILING REGISTER
	PIPE TURNING DOWN	CG	CEILING GRILLE
	UNION	FR	FLOOR REGISTER
	REDUCER - CONCENTRIC	FG	FLOOR GRILLE
	REDUCER - ECCENTRIC	CD	CEILING DIFFUSER
	STRAINER	TV	TURNING VANES
	GATE VALVE	AE	AIR EXTRACTOR
	GLOBE VALVE	SD	SPLITTER DAMPER
	VALVE IN RISER	MD	MANUAL DAMPER
	CHECK VALVE	FD	FIRE DAMPER
	PRESSURE REDUCING VALVE	DL	DUCT LINER IN DUCT
	PRESSURE RELIEF VALVE	AHU	AIR HANDLING UNIT
	SQUARE HEAD COCK	BU	BLOWER UNIT
	BALANCING VALVE	FCU	FAN COIL UNIT
	3-WAY CONTROL VALVE	HWC	HOT WATER CONVECTOR
	2-WAY CONTROL VALVE	UV	UNIT VENTILATOR
	PITCH PIPE MINIMUM 1"/40'	WH	WALL HEATER
	ANCHOR LOCATION	UH	UNIT HEATER
	FLEXIBLE PIPE CONNECTION	WF	WALL FIN RADIATION
	IN-LINE PUMP	PRV	POWER ROOF VENTILATOR
	BOTTOM TAKE-OFF	UVS	UTILITY VENT SET
	TOP TAKE-OFF	PF	PROPELLER FAN
		T	THERMOSTAT
	PRESSURE GAGE	Tₙ	NIGHT THERMOSTAT
	THERMOMETER	Tₕ	THERMOSTAT - HEATING ONLY
	HOT WATER RISER	Tc	THERMOSTAT - COOLING ONLY
	CHILLED WATER RISER		THERMOSTAT - REMOTE BULB
	FAN COIL UNIT	+6'-7"	MOUNTING HEIGHT ABOVE FINISHED FLOOR
	EQUIPMENT AS INDICATED	NIC	NOT IN CONTRACT
	AIR INTO REGISTER		
	AIR OUT OF REGISTER		SUPPLY AIR DUCT SECTION
	AIR FLOW THRU UNDERCUT OR LOUVERED DOOR		RETURN OR EXHAUST DUCT SECTION
	TURNING VANES		FLEXIBLE DUCT CONNECTION
	AIR EXTRACTOR		

Fig. 1-32. Plumbing symbols commonly used on working drawings.

MECHANICAL SYMBOLS

HWS	HOT WATER SUPPLY	PRV	POWER ROOF VENTILATOR
HWR	HOT WATER RETURN	Ⓣ	THERMOSTAT
HSWG	HIGH SIDEWALL GRILLE	MBH	THOUSAND BTU PER HOUR
CD	CEILING DIFFUSER	CFM	CUBIC FEET PER MINUTE
CR	CEILING REGISTER	OA	OUTSIDE AIR
CG	CEILING GRILLE	⟶	AIR OUT OF REGISTER
MD	MANUAL DAMPER	⟶⌄⟶	AIR INTO REGISTER
FD	FIRE DAMPER	⟶Ⓐ⟶	LOUVERED OR UNDERCUT DOOR
UH	UNIT HEATER	⟶⌐	PITCH DOWN AS DIRECTED

Fig. 1-33. Heating, ventilating, and air conditioning symbols currently in use by engineering firms.

points, lines, or surfaces that have a necessary and specific relation to each other or that control the location of other components or mating parts. See Fig. 1-35. Furthermore, dimensions are shown only once on a drawing (so as not to confuse). Only enough measurements are given so that the intended sizes, shapes, and locations can be

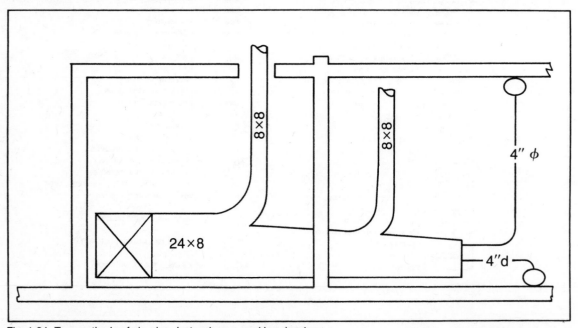

Fig. 1-34. Two methods of showing ductwork on a working drawing.

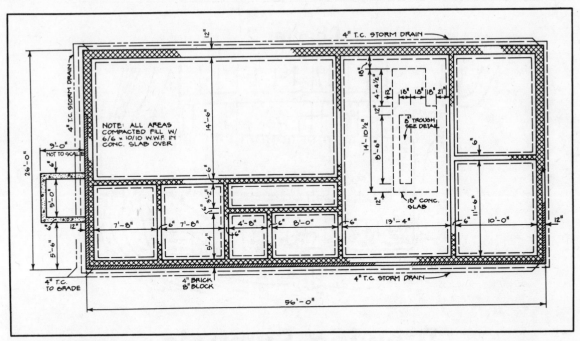

Fig. 1-35. Practical application of dimensioning a working drawing.

determined without assuming any distances.

Once you have gained some experience in reading construction plans and blueprints, reading dimensional lines and figures will become second nature.

Chapter 2

Drawing Symbols

AS MENTIONED EARLIER, THE BASIC PURPOSE of a blueprint or an instructional drawing is to show how a certain project, piece of equipment, or system is to be constructed, installed, modified, or repaired. Such drawings range from those used to assemble a child's toy to intricate drawings used to construct computerized electronic circuits for the space industry. All these drawings have one thing in common. They use symbols and reduced dimensions (in most cases) to convey the required information.

Certain manufacturers sometimes supply pictorial and diagrammatic drawings to help builders better understand how to construct their projects (Fig. 2-1). Heath Company of Benton Harbor, Michigan, always furnishes excellent pictorial drawings with the instructions on how to assemble their many electronic kits. The time required to prepare these drawings is warranted be-

cause the company will sell thousands of these kits using prints or copies of the original. Suppose a fire alarm system is being designed for one small building. Once the system is installed, the drawings will no longer be needed, except perhaps to troubleshoot the system at a later date. If a draftsman was asked to make pictorial drawings of the entire system, chances are the drawings would cost more than the system itself (Fig. 2-2). This type of project will usually use schematic diagrams to convey the necessary information—using various symbols to represent the components of the system.

When available, most designers and draftsmen use symbols adopted by the United States of America Standards Institute (USASI). Many of these symbols are frequently modified to suit a given situation, or new symbols are devised to serve the pur-

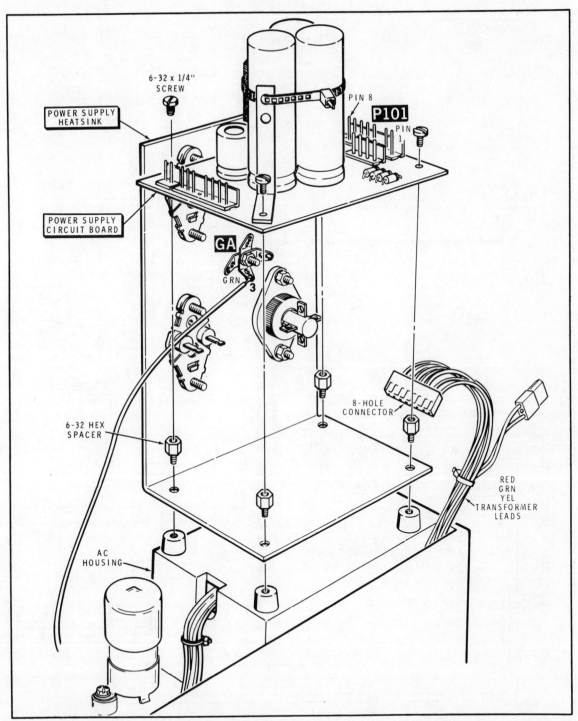

6-32 x 1/4"
SCREW

POWER SUPPLY
HEATSINK

PIN 8

P101

PIN 1

POWER SUPPLY
CIRCUIT BOARD

GA

GRN

3

8-HOLE
CONNECTOR

6-32 HEX
SPACER

RED
GRN
YEL
TRANSFORMER
LEADS

AC
HOUSING

Fig. 2-1. Certain manufacturers use pictorial drawings along with diagrams to aid inexperienced persons in understanding the project (courtesy Heath Company, Benton Harbor, MI 49022).

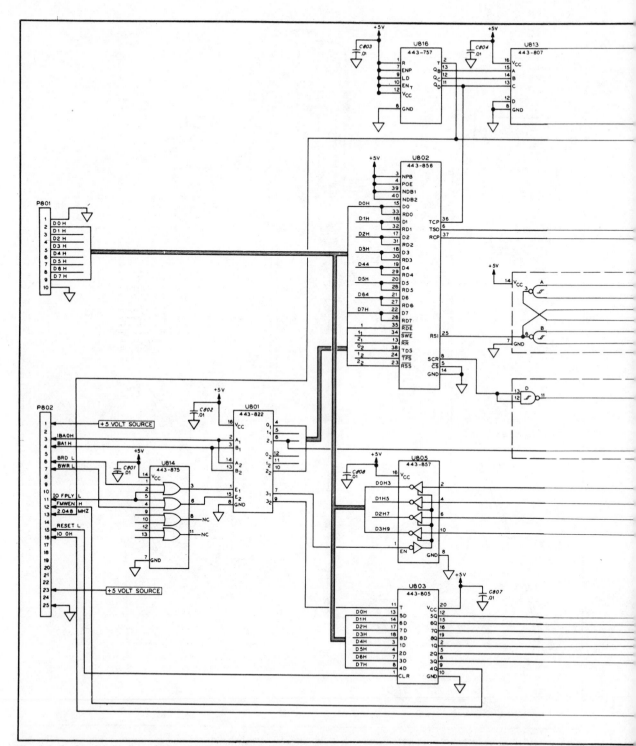

Fig. 2-2. A pictorial drawing of this schematic diagram would probably cost more to make than building the project itself.

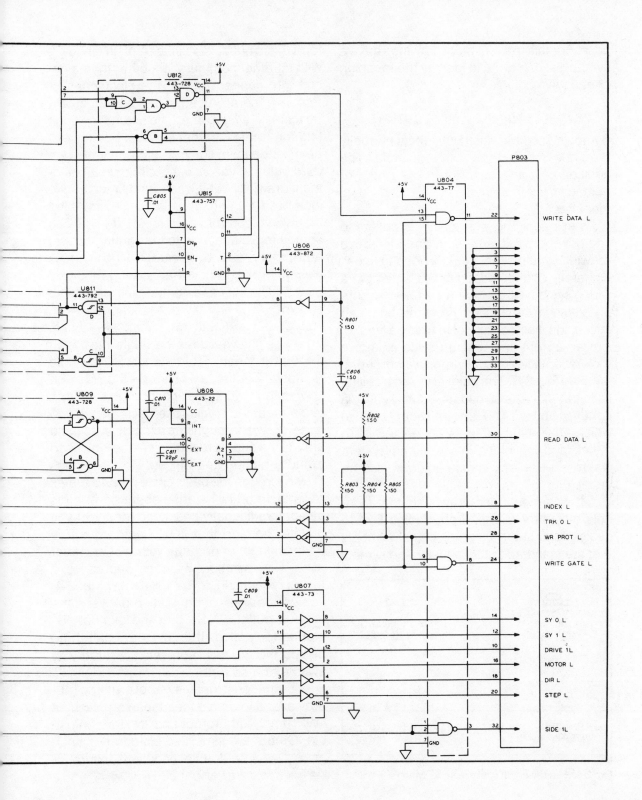

pose. For this reason, most drawings have a symbol list or legend to identify the meaning of each symbol.

SYMBOLS FOR ELECTRICAL OUTLETS

Symbols used for electrical outlets normally have the same basic form—a circle. The addition of a line or a dot to the circle, or perhaps a letter or note, gives it an individual meaning.

Let's continue with electrical outlets to see how simple it is to learn and memorize drawing symbols that you will need to know. Generally an electrical outlet is a point in a wiring system where current is taken to supply electrical appliances. From this brief definition, the electrical outlet to which a lamp or similar current-consuming device is known as a receptacle—usually a *duplex receptacle*. See Fig. 2-3. The point at which a ceiling light is attached is an outlet—usually called a lighting outlet. The larger receptacle to which an electric range is connected is an outlet, and the place in the wiring system where a wall clock is connected is also an outlet.

It was mentioned previously that symbols may vary slightly from design firm to design firm, but in using electrical symbols that are standard in the industry, the symbol

for a basic outlet is a circle. A plain circle (O) with no other configuration, other than a circuit wire connecting it to the electrical system, is generally used to indicate an incandescent lighting fixture. The fixture itself can take on many different shapes, but it is still an incandescent lighting fixture, using standard bulbs. If you saw this circle on an electrical drawing, but it was slightly modified to look like (O⊢), what would be the first thing that would come to your mind? It looks like something is attached to the lighting fixture, and it's a bracket to indicate that the fixture is wall-mounted rather than ceiling-mounted as would be the case if only a plain circle was shown.

Continuing with the circle, ⊖ , this is an outlet, but what does the single line drawn through it mean? This stands for a single receptacle outlet, so only one plug can be connected to it. If one line through the basic circle means a single receptacle outlet, two lines drawn through the circle (⊖) would stand for a duplex receptacle—the kind used in more homes and offices. Do three lines drawn through the basic circle mean a triplex receptacle outlet? It may seem so but this symbol traditionally has stood for a 250-volt outlet—the kind that your electric range is connected to, or perhaps your clothes dryer ⊜ . This symbol ⊜ does represent a triplex receptacle. If you will review Fig. 2-4, you will easily see the relation between the various outlets. Note that the symbol "S" stands for switch. If there is a small three beside the S_3, this represents a three-way switch, and so forth.

Some electrical symbols utilize abbreviations to obtain their meaning, such as "WP" for weatherproof and "N" for nonfusible. Others are simplified pictographs such as ♈ for a double floodlight fixture or ▬▬ for a power panel.

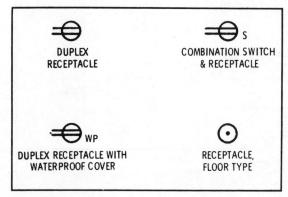

DUPLEX RECEPTACLE

COMBINATION SWITCH & RECEPTACLE

DUPLEX RECEPTACLE WITH WATERPROOF COVER

RECEPTACLE, FLOOR TYPE

Fig. 2-3. Example of various receptacle symbols.

NOTE: THESE ARE STANDARD SYMBOLS AND MAY NOT ALL APPEAR ON THE PROJECT DRAWINGS; HOWEVER, WHEREVER THE SYMBOL ON PROJECT DRAWINGS OCCURS, THE ITEM SHALL BE PROVIDED AND INSTALLED.

FLUORESCENT STRIP

FLUORESCENT FIXTURE

INCANDESCENT FIXTURE, RECESSED

INCANDESCENT FIXTURE, SURFACE OR PENDANT

INCANDESCENT FIXTURE, WALL-MOUNTED

LETTER "E" INSIDE FIXTURES INDICATES CONNECTION TO EMERGENCY LIGHTING CIRCUIT

NOTE: ON FIXTURE SYMBOL, LETTER OUTSIDE DENOTES SWITCH CONTROL.

EXIT LIGHT, SURFACE OR PENDANT

EXIT LIGHT, WALL-MOUNTED

INDICATES FIXTURE TYPE

RECEPTACLE, DUPLEX-GROUNDED

RECEPTACLE, WEATHERPROOF

COMBINATION SWITCH AND RECEPTACLE

RECEPTACLE, FLOOR-TYPE

RECEPTACLE, POLARIZED (POLES AND AMPS INDICATED)

S — SWITCH, SINGLE-POLE

$S_{3,4}$ — SWITCH, THREE-WAY, FOUR-WAY

S_P — SWITCH AND PILOT LIGHT

S_T — SWITCH, TOGGLE W/ THERMAL OVERLOAD PROTECTION

PUSH BUTTON

BUZZER

LIGHT OR POWER PANEL

TELEPHONE CABINET

JUNCTION BOX

DISCONNECT SWITCH - FSS-FUSED SAFETY SWITCH. NFSS-NONFUSED SAFETY SWITCH

STARTER

A.F.F. — ABOVE FINISHED FLOOR

CONDUIT, CONCEALED IN CEILING OR WALL

CONDUIT, CONCEALED IN FLOOR OR WALL

CONDUIT, EXPOSED

FLEXIBLE METALLIC ARMORED CABLE

HOME RUN TO PANEL - NUMBER OF ARROWHEADS INDICATES NUMBER OF CIRCUITS. NOTE: ANY CIRCUIT WITHOUT FURTHER DESIGNATION INDICATES A TWO-WIRE CIRCUIT. FOR A GREATER NUMBER OF WIRES, READ AS FOLLOWS — /// 3 WIRES, //// 4 WIRES, ETC.

— T — TELEPHONE CONDUIT

— TV — TELEVISION — ANTENNA CONDUIT

— S — SOUND-SYSTEM CONDUIT — NUMBER OF CROSSMARKS INDICATES NUMBER OF PAIRS OF CONDUCTORS.

F — FAN COIL-UNIT CONNECTION

MOTOR CONNECTION

M.H. — MOUNTING HEIGHT

F — FIRE-ALARM STRIKING STATION

G — FIRE-ALARM GONG

D — FIRE DETECTOR

SD — SMOKE DETECTOR

B — PROGRAM BELL

Y — YARD GONG

C — CLOCK

M — MICROPHONE, WALL-MOUNTED

M — MICROPHONE, FLOOR-MOUNTED

SPEAKER, WALL-MOUNTED

S — SPEAKER, RECESSED

V — VOLUME CONTROL

TELEPHONE OUTLET, WALL

TELEPHONE OUTLET, FLOOR

TELEVISION OUTLET

Fig. 2-4. Electrical symbol list.

In some cases the symbols are combinations of abbreviations and pictographs such as [⊡] for fusible safety switch and ⊸wp for a weatherproof duplex receptacle.

ELECTRONIC SYMBOLS

Most electronic drawings consist of schematic diagrams utilizing electronic symbols. These symbols are simplified pictorial drawings to represent the various components utilized in electronic equipment. One of the simplest of these is the symbol for a single-cell battery (Fig. 2-5).

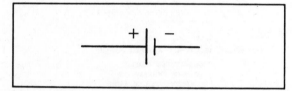

Fig. 2-5. Symbol for single-cell battery.

The longer of the two lines comprising this symbol always represents the positive terminal of the battery, while the short line represents the negative terminal. When this symbol is used in your own diagrams, the short line should be about half the length of the long one.

Amplifiers

Several amplifiers are currently in use, but the basic symbol for all of them is a triangle (Fig. 2-6). Generally an amplifier is used to produce an amplified reproduction of its

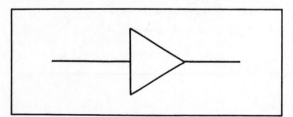

Fig. 2-6. Electronic amplifier.

input signal by drawing power from a source other than the input signal. Some amplifier letter combinations that may be used with amplifier symbols, and which you are certain to run across, are listed below.

BDG	bridging
BST	booster
CMP	compression
DC	direct current
EXP	expansion
LIM	limiting
MON	monitoring
PGM	program
PRE	preliminary
PWR	power
TRQ	torque

Antennas and Lightning Arresters

Antennas are used on radio, television and other signal receivers for picking up signal waves (Fig. 2-7). The antenna itself usually consists of a metallic structure or an arrangement of conducting wires or rods.

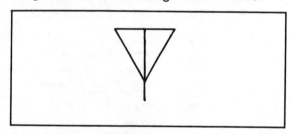

Fig. 2-7. Symbol for antennas.

Several symbols have been used for *lightning arresters* through the years, and those who must read electronic drawings should be familiar with them all (Fig. 2-8). You will see carbon block, electrolytic, horn gap, protective gap, and others. All are protective devices used to sidetrack or divert a discharge of lightning directly to ground without damaging an electronic device or its transmitting or receiving antenna.

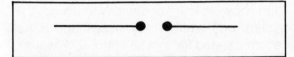

Fig. 2-8. Symbol for lightning arrester.

Audible Signaling Devices

Audible signaling devices consist of bells, buzzers, horns, loudspeakers, and the like (Fig. 2-9). All are capable of being heard by the human ear. This means that any audible signaling device must be within 20 to 20,000 cycles—the range of the human ear.

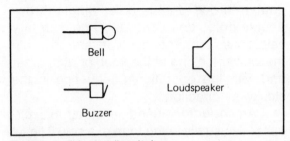

Fig. 2-9. Audible signaling devices.

Loudspeaker letter combinations used with symbols on electronic drawings include:

HN	electrical horn
HW	howler
LS	loudspeaker
SN	siren
EM	electromagnetic with moving coil
EMN	electromagnetic with moving coil and neutralized winding
MG	magnetic armature
PM	permanent magnet

Batteries and Capacitors (more than one)

A *battery* consists of dry cell or storage cell connected to serve as a dc voltage source. A single cell is not a battery.

A *capacitor* is used in electronic circuits to block the flow of direct current while allowing alternating current and pulsating cur-

rent to pass. The device consists of two conducting surfaces separated from each other by an insulating material such as air, oil, paper, glass, or mica. A capacitor is capable of storing electrical energy. The curved line used in the symbol represents the outside electrode in fixed-paper and ceramindielectric capacitors, the moving element in variable and adjustable types, and the low-potential element in feed-through capacitors.

There are other symbols used to indicate capacitors that do not appear in some lists of standards. If you are involved in drawing, you should be on the alert for such symbols. Other symbols related to the general capacitor symbol include the variable capacitor, the adjustable capacitor with mechanical linkage, the continuously adjustable differential, the phase shifter, the split stator, and the feed-through capacitor.

Photoelectric Cells and Circuit Breakers

The *photoelectric cell* is a type of light-sensitive cell in which a treated cathode, mounted in a glass envelope, emits electrons under the action of light. These electrons are collected by the anode. Another type varies its resistance with the change of light falling on the cell.

A *circuit breaker* is an electromagnetic device that opens or breaks a circuit automatically when the current rises above a predetermined value. These devices are rapidly replacing fuses in almost all electronic devices and instruments.

Rectangles

A rectangle is often used on electronic drawings to represent various circuit elements. To indicate different components, letter combinations are placed either inside or adjacent to the basic symbol. The ones to follow are typical of those used.

CB	circuit breaker
DIAL	telephone dial
EQ	equalizer
FAX	facsimile set
FL	filter
FL-BE	filter, band elimination
NET	network
PS	power supply
RU	reproducing unit
RG	recording unit
TEL	telephone station
TPR	teleprinter
TTY	teletype

Ground and Coil

The connection to earth, which is made by a buried conductor, is called a *ground*. Also, the chassis of an electronic instrument is called a ground and serves as the return path for signals. Ground is usually considered to be at zero r.f. (reactive factor) potential.

A *coil* is a number of turns of wire wound on an iron core or on a coil form made of insulating material. A coil offers considerable opposition to the passage of alternating current but very little opposition to direct current.

Connectors and Connections

Any device that joins or couples two or more parts is called a *connector*. Note in the symbol lists that female and male connectors are shown separately.

A *connection* is any wire or other conductor that joins another wire or component in an electronic circuit. Two systems are approved for showing connections. One is the dot system (this is the most accepted method). The other is using no other symbol

than showing lines or devices touching; this system has some weaknesses and is not recommended for most electronic drawings. When the dot system is used to show connections, crossovers are shown by merely crossing the wires, as would be shown for a no-dot connection.

Switches and Inductors

The purpose of a *switch* is to open and close an electrical circuit or to change the connections between components or circuits. Some types are the single-pole; single-pole, double-throw; double-pole, single-throw; double-pole, double-throw; and a rotary selector switch. The last type is drawn with the viewer looking at the knob or at the rear and with the operational sequence in the clockwise direction.

An *inductor*—used in electronic circuits—is usually in the form of a coil. It limits the flow of alternating current while allowing direct current to pass. When an inductor is used in this manner, it is usually referred to as a choke coil. Radio-frequency choke coils have air or pulverized iron cores, while audio-frequency choke coils and filter chokes have iron cores.

Resistors

An electronic component that offers resistance to the flow of electric current is called a *resistor*. Its resistance is specified in ohms or megohms, and its power rating is in watts. Because resistors offer a resistance to the flow of current, they are used to drop the voltage to the correct value for operation of transistors. They also isolate circuit components from each other, so that one will not interfere with the operation or action of the other.

The zigzag symbol—made with a 60-

degree angle between adjacent lines—is probably the symbol that is used most on most electronic drawings. Rectangles are also used in some drawings with the value of the resistor placed inside the rectangle or near it.

Resistors may be fixed, variable, or tapped. See Fig. 2-10.

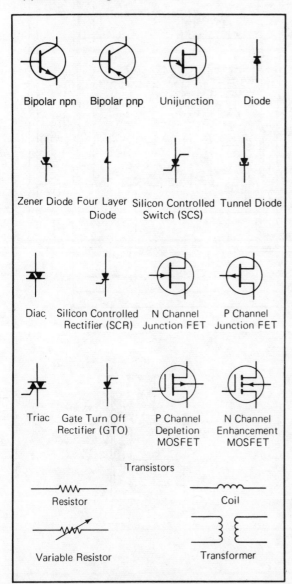

Fig. 2-10. Electronic drawing symbols.

Relays, Transformers, and Loudspeakers

A *relay* may be used to control the current in one circuit by a much smaller current flowing in another circuit. It is an electromagnetic device with either normally open or normally closed contacts. These contacts act as switches and are turned on and off by action of the relay coils.

A *transformer* is a device consisting of two or more coils constructed such that magnetic lines of force produced by the flow of alternating or pulsating direct current through one coil will pass through the other coil and induce in it a corresponding alternating voltage. The symbols used for inductors are also used for drawing transformers on electronic drawings.

A *loudspeaker* is a device used to radiate acoustic power effectively at a distance through the air. When incorporated into radios, televisions etc., it usually reproduces and magnifies the tone from the oscillator. A pictorial symbol is normally used to show a speaker on an electronic drawing. Templates will aid the draftsman in drawing this and other similar symbols.

Transistors

A *pnp transistor* is one in which the base is made up of semiconductor material that has been doped with a donor material that gives an excess of electrons. The emitter and collector are made of semiconductor materials that have been doped with an acceptor material that gives an excess of holes.

A *npn transistor* is one in which the material arrangements are in reverse of a pnp transistor. Both types are best drawn on electronic drawings by using readily available templates.

Rectifiers and Conductors

A *rectifier* is a device that changes an alter-

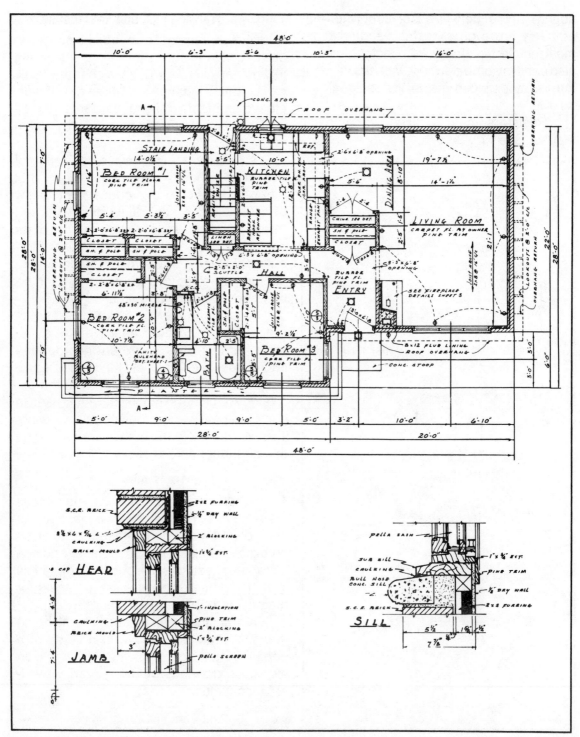

Fig. 2-11. Plan and sectional drawing utilizing, in part, some of the available symbols.

nating current into a pulsating direct current.

The common grouping of conductors may be indicated on electronic drawings in several ways. Multiconductor cables are shown by encircling the conductors with a loop. A broken line indicates that the cable is shielded.

ARCHITECTURAL SYMBOLS

Most other architectural symbols are pic-

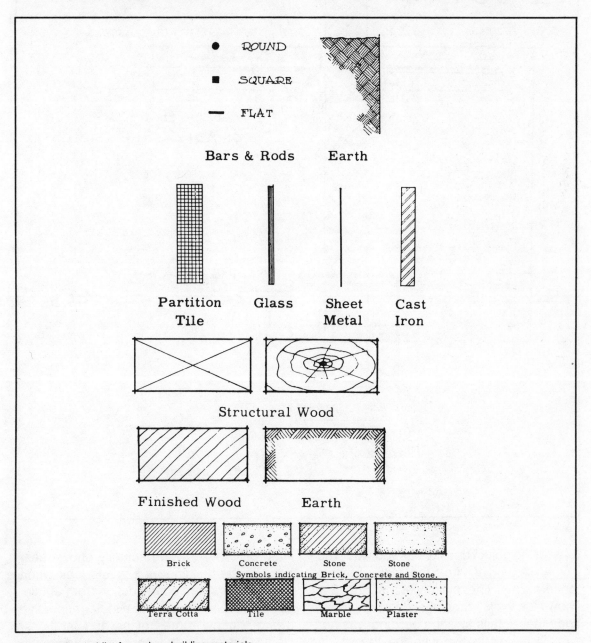

Fig. 2-12. Symbol list for various building materials.

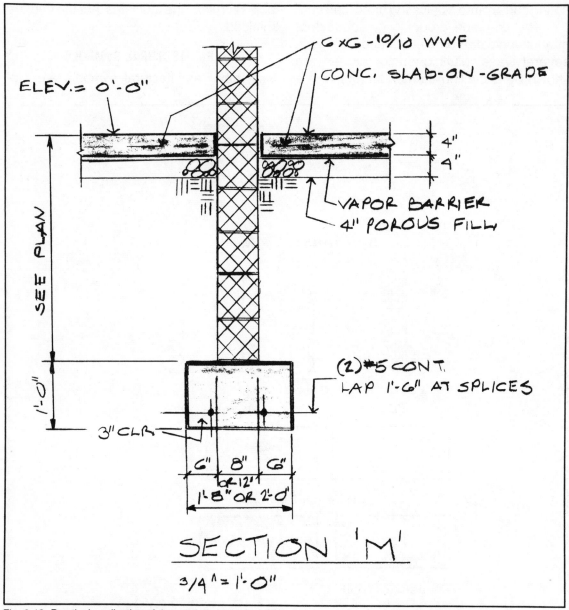

ELEV.= 0'-0"

6×6 -10/10 WWF

CONC. SLAB-ON-GRADE

4"
4"

SEE PLAN

VAPOR BARRIER
4" POROUS FILL

(2) #5 CONT.
LAP 1'-6" AT SPLICES

1'-0"

3" CLR.

6"   8"   6"
OR 12"
1'-8" OR 2'-0"

SECTION 'M'
3/4" = 1'-0"

Fig. 2-13. Practical application of the symbol shown in 1-20.

tures of the object depicted as you would see it when looking at it from the angle shown on the drawing. The plain view in Fig. 2-11 shows the walls, partitions, doors, windows, and other details as if the building was horizontally sliced in two with the top section lifted off. The drawing actually shows sections of the building as it would look when viewing the sliced open section from above. All doors are shown with their respective swing, window frames, glass, and partitions. Using Fig. 2-12, the persons reading the

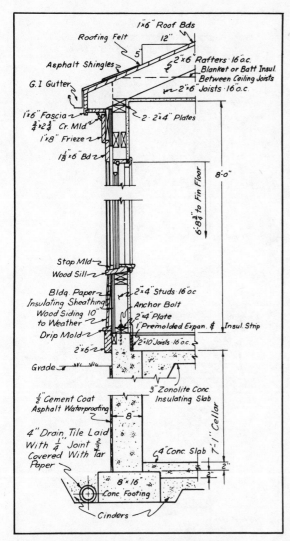

Fig. 2-14. Practical use for symbols in Fig. 1-22.

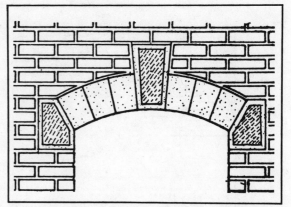

Fig. 2-15. Practical applications of the symbols in Fig. 1-26.

drawing can tell with what material each section of the building is constructed. When symbols are not appropriate, notes are used to indicate the finish or material.

SYMBOLS FOR VARIOUS MATERIALS

Figure 1-20 shows the three most common symbols used to indicate earth, rock, and stone fill. An example of practical use is shown in Fig. 2-13. Notice that the earth area

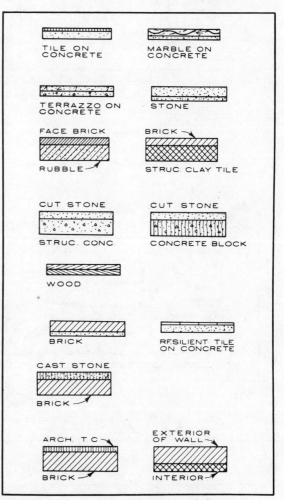

Fig. 2-16. Miscellaneous architectural symbols.

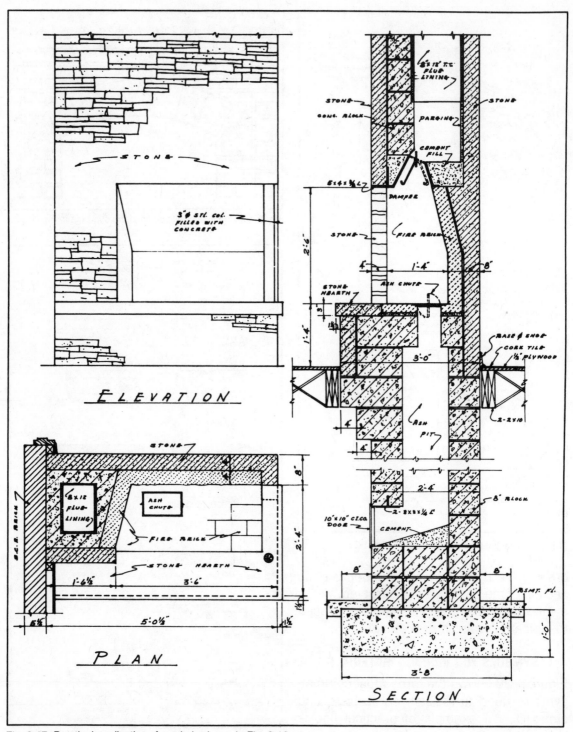

Fig. 2-17. Practical application of symbols shown in Fig. 2-16.

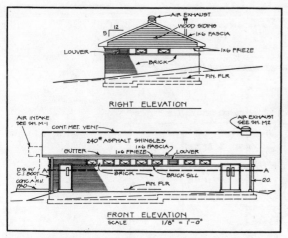

Fig. 2-18. Elevation of a building showing the application of various architectural symbols to indicate finishes.

is not entirely marked with the earth symbol. Rather, only the areas around other material are indicated by the symbol.

Symbols for concrete and related materials such as concrete block, terrazzo, etc., are shown in Fig. 1-22. Examples of their use appear in Fig. 2-14.

Several symbols are used to indicate metal. The type employed is dependent upon the size of the drawing or the scale to which the drawing is drawn. When the scale used is too small for the symbol, notes are normally provided near the items indicating the type of metal to use. Some of these symbols are shown in Fig. 1-24.

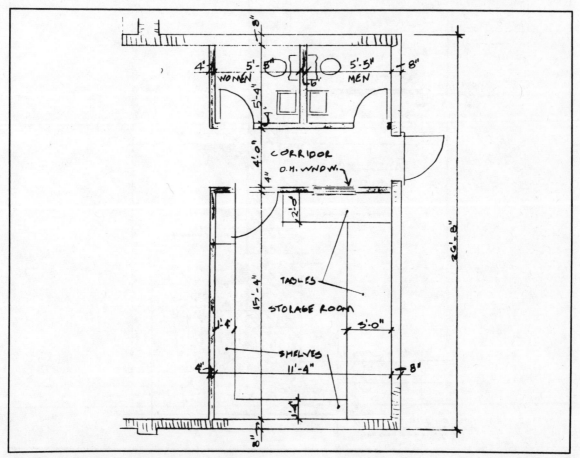

Fig. 2-19. Plan views of exterior walls of a building.

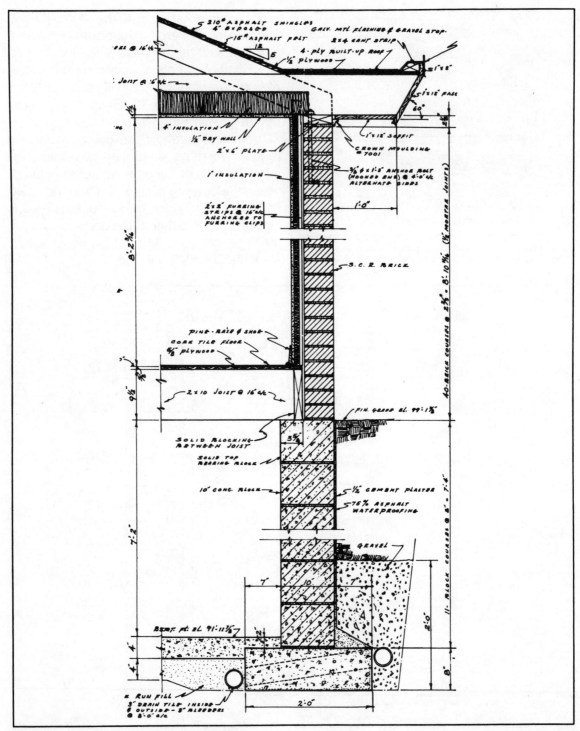

Fig. 2-20. Section of a building showing floor finish.

34

Wood symbols appear in Fig. 1-25. As for the metal symbols, the size of the drawing determines the type of symbol used.

Symbols commonly used to indicate stone and brick are shown in Fig. 1-26. Practical applications of some of the symbols are shown in Fig. 2-15.

Figure 2-16 shows an assortment of miscellaneous architectural symbols. Their practical applications are in Fig. 2-17.

Symbols used in building elevation drawings are shown in Fig. 2-18. Previously described symbols are used when the need arises. Other material combinations are shown in Figs. 2-19 (plan views of exterior walls) and 2-20 (sections of floor finish).

PLUMBING SYMBOLS

Most plumbing drawings utilize lines and symbols to signify pipe, fittings, fixtures, and the like. A typical set of plumbing fixtures appears in Fig. 1-32. Using these symbols makes the draftsman's job easier and also helps the workmen on the job to understand the intent of the design.

HEATING, VENTILATING, AND AIR CONDITIONING (HVAC) SYMBOLS

The main purpose of HVAC drawings is to show the layout of the heating and cooling units, the ductwork, and piping systems. Graphic symbols on HVAC drawings are not unlike those used in plumbing. The symbols shown in Fig. 2-21 are typical of those in current use.

WELDING SYMBOLS

The American Welding Society recommends the welding symbols shown in Fig. 2-22 for use on working drawings. These symbols may vary on some drawings, but if they do, usually a legend will be provided to indicate their meaning. If you are required to read structural drawings or other types that need welding, you should be familiar with these symbols.

RIVET SYMBOLS

If you are involved with metal working drawings of any type, you should become familiar with symbols used to indicate the various types of rivets. Some are shown in Fig. 2-23. Dimensions of rivets are normally indicated by notes and numerals. See Chapter 14.

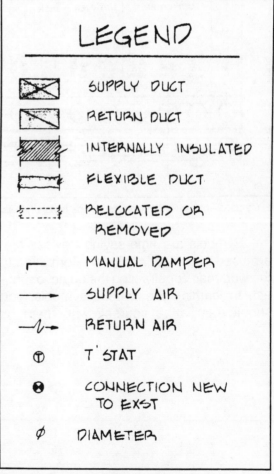

Fig. 2-21. Plumbing symbols in current use.

Welding Symbols									
Bead	Fillet	Plug or slot	Groove					Contour	
			Square	V	Bevel	U	J	Flush	Convex
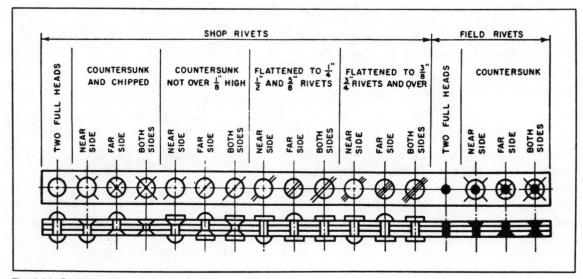									

Fig. 2-22. Welding symbols recommended by the American Welding Society.

Fig. 2-23. Symbols used to indicate various types of rivets.

Symbols are time-saving devices that are used to convey pertinent information to the workman constructing the object or project. In learning the various symbols, you should familiarize yourself with them by glancing over the lists several times. Don't be concerned if you can't memorize all of them at this time. As you gain experience, you will retain the meaning of each automatically.

36

Chapter 3

Schedules and Their Use

A *SCHEDULE,* AS APPLIED TO BLUEPRINT READING, is a systematic method of presenting notes or materials lists, equipment, components, and the like on a drawing in tabular form. When properly organized and thoroughly understood by both the draftsman and those reading the drawings, schedules are great time-saving devices.

The checklist schedule for an electronic project is shown in Table 3-1. This schedule allows the person building the project to perform tests to assure that correct connections have been made. The schedule is to be used with a voltmeter. Note that the person testing the circuit reads the top line of the schedule; he places the negative test lead to P102, pin 2 and the positive test lead to P102, pin 1. The voltmeter reading should show 70 volts direct current. If not, something in the circuit is at fault.

If the same information contained in this schedule was included in the written specifications, more time would be required to explain how the test is made. The person building the project would have to comb through pages of written specifications to find the explanation.

The following schedules are typical of those used on all construction drawings. While the list is by no means complete, it is a good sampling of those used for the various fields utilizing blueprints and working drawing. Study the schedules until you are thoroughly familiar with their applications.

SEMICONDUCTOR SCHEDULE

Table 3-2 was used on a how-to project to identify the various components given in a schematic diagram. The identification number is listed on the schematic view. The schedule then tells what the part is (by catalog number) and the quantity of each

Table 3-1. Checklist Schedule for an Electronic Kit Project.

	NEGATIVE LEAD TO:	POSITIVE LEAD TO:	APPROXIMATE METER READING:
( )	P102, pin 2	P102, pin 1	70 VDC
( )	P102, pin 4	P102, pin 3	13 VDC
( )	P102, pin 7	P102, pin 6	+20 VDC
( )	P102, pin 7	P102, pin 8	−20 VDC
( )	Heat sink	S516 pin 3 (red wire)	5 VDC
( )	Heat sink	S516 pin 1 (orange wire)	+20 VDC

part needed to complete the project. Most component schedules found on electronic drawings will be similar to this type—that is, materials lists and where they are to be used in a particular circuit.

LIGHTING FIXTURE SCHEDULE

The lighting fixture schedule shown in Table 3-3 lists the fixture type and identifies each fixture type on the drawings for a given project by number. The manufacturer and identification number of each type are given, along with the number, size, and type of the lamps used for each.

FINISH SCHEDULE

Table 3-4 is typical of those schedules used on architectural drawings to indicate the finishes of rooms or areas in buildings. The extreme left-hand column with the heading "Mark" gives the room number. These marks also appear on the floor plans of the building in question. The remaining columns cover the finishes of the floor, base, wainscot, walls, and ceiling.

Note the references to notes such as, "See Note #1, #2, #3, etc." On this project a set of notes accompanied the finishing schedule and are as follows:

● Interior walls are of stone to 7 feet, 7½ inches above the floor line. See Details A-2.

● Match existing plaster finish and wood trim in all areas where the existing is disturbed or new work abuts the existing.

● Exterior wall finish to be applied over 1½ inch rigid insulation and furring channels. Interior wall finish (over masonry) is to be applied over 1 × 2 channels or furring strips. The furring is to be 2 × 4 at the existing stone wall.

Table 3-2. A Table Used for an Electronic Project to Identify the Various Components.

DIODES — TRANSISTORS — INTEGRATED CIRCUITS (IC's)

C1	56-56	3	1N4149 diode	D201, D204, D207
C1	56-58	1	6.2 V zener diode	D205
C1	56-73	1	MZ2360 diode	D202
C1	56-93	1	FD333 diode	D206
C1	56-94	1	12.8 V zener diode	D203
C1	57-27	4	1N2071 diode	D209, D212, D214, D215
C1	57-64	1	DRS110 diode	D211
C1	57-614	1	MR508 diode	D208

NOTE: Transistors (and integrated circuits) are marked for identification in one of the following four ways:

1. Part Number.
2. Type number.
3. Part number and type number.
4. Part number with a type number other than the one shown.

C2	417-811	1	MPSL01 transistor	Q209
C2	417-821	1	MPSA06 transistor	Q204
C2	417-822	1	MPSA56 transistor	Q203
C2	417-823	1	MPU131 transistor	Q201
C2	417-874	1	2N3906 transistor	Q212
C2	417-875	1	2N3904 transistor	Q215
C2	417-885	1	MPSA65 transistor	Q202
C3	417-195	1	MJE340 transistor	Q216
C3	417-924	2	MJE172 transistor	Q208, Q211
C3	417-932	1	MJE182 transistor	Q207
C4	417-834	1	MPSU10 transistor	Q214
C4	417-926	2	MPSU06 or NSDU06 or RCP701C transistor	Q205, Q206
C5	417-282	1	MJ2841 transistor	Q213
C5	417-923	1	BU500 transistor	Q217
C6	442-53	2	NE555 integrated circuit	U201, U202

Table 3-3. A Lighting Fixture Schedule. It is a Great Time-Saving Device When Used on Working Drawings for Electrical Installations. Without Such a Schedule, a Complete Specification for Each Lighting Fixture Would Have to be Provided.

LIGHTING FIXTURE SCHEDULE

TYPE	MANUFACTURER	MODEL NO.	VOLT-AGE	WATTS LAMPS	MOUNTING	REMARKS
A	Holophane	SG4Q1-PB.-FN82QS	120	200 /4- F40CW	Recessed	Inboard & outboard lamps switched separately
B	Holophane	7400-4-120	120	200 /4-F40CW	Pendant	8'-0" Abv. finished floor
C	Marco	C437024	120	120 /2-60W	Surface	
D	Marco	C738024	120	300 /4-75W	Surface	
E	Marco	C7337024	120	60 /1-60W	Surface	
F	See DWG M-2	—	120	/1-100W	Recessed	
G	Holophane	6100-4-120	120	50 /1-F400W	Surface	
H	Devine Ltg. Inc.	B306-C85	120	175W /MV-H39-BT28	Wall Surface	Line up top of fixt. with top of windows, photo-electric cell (C85)
I	Crescent Ltg. Corp.	ANG220-H	120	50 /2-F20CW	Wall Surface	Above mirror
K	—	Pit light	120	60 /1-60W	Surface	Size fixture & locate to clear elevator
⊗	Holophane	M-22	120	25 /1-F20T12	Wall 6" abv. door	Battery operated upon power failure.
⬠	Holophane	M-6-bronze-120	120 / 6	110 /2-8W sealed beam	Wall 6'-6" AFF	Battery operated upon power failure.

Table 3-4. Working Drawings for Building Construction Frequently Use a Finish Schedule to Indicate Finishes of Rooms or Areas Within the Building.

MARK	DESIGNATION	FLOOR		BASE		WAINSCOT		FINISH SCHEDULE
		MAT.	COLOR	MAT.	COLOR	MAT.	COLOR	
101	Vestibule	V. asb. tile		—		—		
102	Stair hall (existing)	Exist.		Exist.		Exist.		
103	Clerks's conference room (existing)	Exist.		Exist. see note #2		—		
104	Commissioner of accounts (existing)	Exist.		Exist. see note # 2		—		
105	Clerk's office general (existing)	Exist.		Exist. see note # 2		—		
106	Record room (existing)	Exist.		Exist.		—		
107	Micro-film	V. asb. title		4" vinyl		—		
108	Elevator machinery	Conc.		—		—		
109	Record room	V. asb. tile		4" vinyl		—		
110	Security storage	V. asb. tile		4" vinyl		—		
111	Men's toilet (existing)	Exist.		Exist.		Exist.		

MARK	WALLS		CEILING			REMARKS
	MAT.	COLOR	MAT.	COLOR	HGT.	
101	Stone		Painted ½" Gyp. bd. suspended		7'-7½"	See note #1
102	Exist.- see note #1		Exist.		—	—
103	Exist. see note #2		Exist.		—	Match exist. wd. base where door is added. Match finish where door is added.
104	Exist. see note #2		Exist.		—	Match Exist. wd. base where door is added.
105	Exist. see note #2		Exist.		—	Match finish where door is added.
106	Exist.		Exist. struct.		—	
107	Painted C.M.U.		Ac. tile		9'-0"	
108	C.M.U.		Conc.		10'-10"±	
109	Painted ½" gyp. bd.		Painted conc.		10'-10"±	See note # 3
110	Painted C.M.U.		Painted conc.		10'-10"±	
111	Exist.		Exist.			See plan, sht. 4-exist. finishes disturbed shall be patched to match exist. space shall be re-painted

• New vinyl asbestos floor tile is to be applied to match the tile selected for Space 201.

• Provide a new wood base for this area to match the existing.

• Provide a new wood base on new partitions to match the existing.

• Provide a new chair rail to match existing.

• Provide a new chair rail on new partitions to match the existing. Wainscot wall finish shall match the existing.

• Wall finish shall match the existing.

• Remove existing floor finish and furnish new vinyl asbestos tile. Wall finish and wood trim shall match the existing. Existing walls will be painted to match the new work.

• Remove the existing floor finish from removed partitions to the new partition at Space 208 and furnish new vinyl asbestos tile. Wall finish and wood trim shall match the existing. Existing walls will be painted to match the new work.

• Wall finish and wood trim shall match the existing. Existing walls will be painted to match the new work.

• Floor plates for partitions between corridor and offices and for the partition between offices shall be set in two beads of caulking. Partitions shall be insulated with

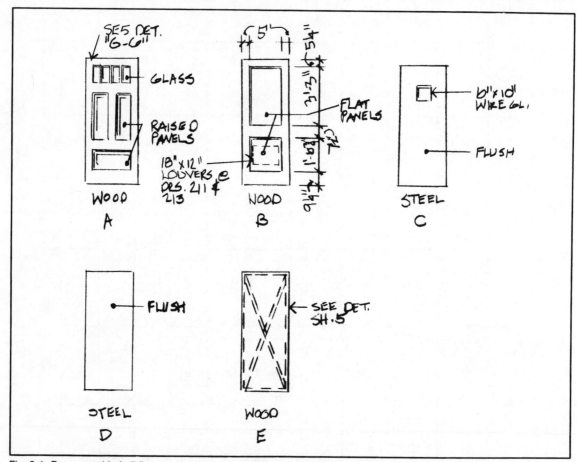

Fig. 3-1. Doors used in building construction projects are sometimes drawn to scale along with all details.

Table 3-5. Door Schedules Are Used on Working Drawings to Indicate Details, Sizes, and Finishes on Each Door Used on the Project.

DOOR SCHEDULE

MARK	TYPE	SIZE WIDE	SIZE HIGH	THK.	FRAME TYPE	LINTEL	T'HOLD	REMARKS	H'DWARE NO
101	A	3'-0"	7'-0"	1¾"	Det. "G-6"	Steel	Bronze	See details "G & M-6"	1
102	Exist.	—	—	—	Det. "J-6"	Exist.	Exist.	See details sht. 6-change door to swing out.	7
103	B. exist.	2'-8"	7'-0"	1¾"	Sim. to det. "D-6"	Wd.	—	Re-use door from sp. 103-verify size	—
104	C	2'-8"	6'-8"	1¾"	Det. "H-6"	Exist.	—	Note 4" head	11
105	C	2'-8"	6'-8"	1¾"	Det. "H-6"	Exist.	—	Note 4" head	3
106	C.O.	—	—	—	Det. "A-6"	Exist.	—	See details sht. 6	—
107	C.O.	—	—	—	Det. "A-6"	Exist.	—	See details sht. 6	—
108	D	2'-8"	6'-8"	1¾"	Det. "F-6"	CMU	—		5
109	D	2'-10"	6'-8"	1¾"	Det. "F-6"	CMU	—		3
110	D	2'-8"	6'-8"	1¾"	Det. "F-6"	CMU	—		5
111	E	2'-8"	6'-6"	3½"	See details	Wood	—	See details sht. 5	19
112	B	2'-10"	7'-0"	1¾"	Sim. to det. "D-6"	Exist.-verify w/ new wider opng.	—	See ¼" scale plan @ sht. 4	9

3½-inch fiberglass batt insulation. Partitions shall have sound deadening material as indicated on the plans.

● Match adjacent existing wall finishes and wood trim at all new openings or work abutting the existing.

DOOR SCHEDULE

Some architects like to show the door types used on a project (Fig. 3-1). There is usually a door schedule accompanying the door types to give details concerning the sizes and finishes of each type. See Table 3-5. As with the finish schedule, the "Mark" column gives the room or area number which is also used on the floor plan of the construction drawings. The type—A, B, C, D, or E— matches those appearing in Fig. 3-1. The remaining information gives dimensions of each door used along with frame and lintel types, etc. Some doors also have remarks. The hardware used on each door is indicated in the extreme right-hand column. The numbers correspond to those used in the written specifications to exactly describe each type of hardware.

WINDOW SCHEDULE

The window schedule in Table 3-6 is typical of those used on almost every architectural drawing to indicate the type, size, and other necessary details of the windows used in the building.

GRILLE AND DIFFUSER SCHEDULE

A typical grille and diffuser schedule such as the one shown in Table 3-7 should indicate the manufacturer and catalog number of each grille and diffuser used on a given project. Further information should include the dimensions of each and the volume of air in cubic feet per minute (CFM) that each will properly handle. A column for remarks is included to offer more information that will be helpful in installing the grille or diffuser.

DIFFUSER SCHEDULE

Another type of diffuser schedule is shown in Table 3-8. This schedule was used on a set of construction drawings for a large residence. The extreme left-hand column indicates the room or area in which the different

Table 3-6. A Window Schedule Is Similar to a Door Schedule and Is Used in the Same Way—to Give Details Concerning Sizes and Finishes.

WINDOW SCHEDULE						
MARK	TYPE	SASH SIZE	MASONRY OPENING		LINTEL	REMARKS
			INTERIOR	EXTERIOR		
A	Plaster niche			4'-0"×7'-5"	Steel	See section "A-7"
B	Wood double hung	3'-10¾"×6'-4⅝"	4'-8"×6'-11⅛"	4'-8"×7'-3⅝"	Steel & CMU	See section "C-7" & Det K-6"
C	Wood double hung	3'-0¾"×5'-7⅝"	3'-10"×6'-3⅛"	3'-10"×6'-11½"	Steel & CMU	See section "C-7" & Det. "L-6"

Table 3-7. Grille and Diffuser Schedules Appear on Most HVAC Drawings.

	GRILLE		SCHEDULE		
MARK	MFGR	MODEL	SIZE	CFM	REMARKS
1	KRUEGER	SHI0	9 x 9	146	OBD ALUMINUM
2		SH 3	6 x 6	50	
3				60	
4				75	
5				80	
6				100	
7			9 x 9	144	
8				152	
9				170	
10				175	
11				181	
12				226	
13		SH4	6 x 6	65	
14				100	
15			9 x 9	175	
16				222	
17		S580H	8 x 6	150	
*18			14 x 8	292	
19				300	
*20			18 x 8	354	
21			24 x 8	500	
22		S580V	34 x 12	1000	
23		SH 3	9 x 9	205	
24	KRUEGER	S580V	8 x 6	167	C.R OBD ALUM
25		S580V	18 x 8	362	C.R OBD ALUM

*INCLUDE FILTER FRAME 5FF HINGED

diffusers are to be located during construction. The next column, to the immediate right, indicates the quantity, manufacturer's name, model number, and dimensions of the supply air diffusers for each area. The next column indicates the air volume in cubic feet per minute. The last column is for any necessary remarks.

EXHAUST FAN SCHEDULE

The schedule shown in Table 3-9 was used on a set of heating, ventilating, and air conditioning drawings for a residential building. Note that the left-hand column gives the room or area designation. The remaining column indicates the manufacturer, model number, cubic feet per minute, and the sound levels in decibels of the various fans used. The "head" output from the combination heat and fan units in the laundry room and master bath, as well as the controls, is listed in British thermal units per hour.

HOT-WATER BOILER SCHEDULE

The schedule appearing in Table 3-10 was used on an actual construction project to save time for the draftsmen and workers. If such a schedule had not been used, a complete description of this boiler would have been necessary. It would have been worded similar to the following:

The HVAC contractor shall furnish and install—as indicated on the drawings—one oil-fired boiler having a firing rate of 2.55 gallons per hour (gph) and an I.B.R. gross output of 261,000 Btu/h. The boiler shall be constructed of cast iron in accordance with ASME requirements for low-pressure heating boilers and bear the ASME symbol. Each section shall be factory tested at 2½ times maximum working pressure of 50 pounds for water. The boiler shall have an I.B.R. rating.

The boiler shall be composed of cast-iron sections with vertical flutes. Manufacturers will furnish boilers with an aluminized steel canopy. Water or steam trim as required. Oil burners shall include controls as described herein.

The boiler shall be equipped with a tankless heater having a rating of 7 gallons per minute (gpm).

RADIATION SCHEDULE

Hot-water or steam baseboard radiators may be described in a schedule such as Table 3-11. Note the following:

● Symbol R/1, as indicated in the schedule, means that this line (in the schedule) describes radiator number 1,

Table 3-8. Another Type of Diffuser Schedule.

SUPPLY DIFFUSER SCHEDULE LOWER LEVEL			
GUEST B.R.	2 - CARNES MODEL 7292-A x 4'0" LONG	157 - CFM EACH	
GAME RM.	BASE BOARD ELECTRIC		
BATH #1	I - CARNES MODEL 200 - 12 x 4 REG.	NO VOLUME CONTROL 50 - CFM EACH	STACK HEAD DAMPER
BAR	I - CARNES MODEL 200 - 12 x 4 REG.	NO VOLUME CONTROL 121 - CFM EACH	STACK HEAD DAMPER
KITCHENETTE	I - CARNES MODEL 200 - 12 x 4 REG.	NO VOLUME CONTROL 64 - CFM EACH	STACK HEAD DAMPER
GALLERY	I - CARNES MODEL C-40 FLOOR 3" x 5'0" 2 - CARNES MODEL 7261-A x 3'0" LONG	200 - CFM EACH 100 - CFM EACH	
STUDY	I - CARNES MODEL 7292-A x 6'0" LONG	282 - CFM EACH	
BED RM. #1	I - CARNES MODEL 7261-A x 6'0" LONG	188 - CFM EACH	
BED RM. #2	I - CARNES MODEL 7261-A x 6'0" LONG	188 - CFM EACH	
BATH #2	I - CARNES MODEL 200 - 12 x 4 REG.	NO VOLUME CONTROL 50 - CFM EACH	STACK HEAD DAMPER
LAV. #1	I - CARNES MODEL 200 - 12 x 4 REG.	NO VOLUME CONTROL 50 - CFM EACH	STACK HEAD DAMPER
LAV. #2	I - CARNES MODEL 200 - 12 x 4 REG.	NO VOLUME CONTROL 50 - CFM EACH	STACK HEAD DAMPER
RETURN AIR DIFFUSER SCHEDULE			
GAME RM.	BASE BOARD ELECTRIC		
GUEST B.R.	CARNES MODEL 7295 x 4'0"	250 - CFM EACH	
GALLERY	CARNES MODEL 7295 x 4'0"	400 - CFM EACH	
BED RM. #1	CARNES MODEL 7295 x 4'0"	238 - CFM EACH	
BED RM. #2	CARNES MODEL 7295 x 4'0"	238 - CFM EACH	
STUDY	CARNES MODEL 7295 x 4'0"	332 - CFM EACH	

Table 3-9. An Exhaust Fan Schedule Used on a Set of HVAC Drawings for a Residential Building.

EXHAUST FAN SCHEDULE	
BATH No. I	NUTONE MODEL QT-110-110 CFM DECIBELS SOUND LEVEL 2.5
BATH No. 2	NUTONE MODEL QT-80-80 CFM DECIBELS SOUND LEVEL 1.5
LAY # I	" " " " " " " " "
LAY # 2	" " " " " " " " "
LAUNDRY RM.	2 EA NUTONE HEAT-A-VENT # 9275-4.265 B.TUH HEAT 40 CFM EA
	WITH 2 EA ON/OFF HEAT/VENT CONTROLS
MASTER BATH	I EA NUTONE HEAT-A-VENT # 9275-4.265 B.TUH-HEAT 40 CFM WITH
	ONE EA ON/OFF HEAT CONTROL WALL MOUNTED THERMOSTAT

Table 3-10. This Schedule Was Used on an Actual Construction Project to Save Time for the Draftsman and Workers.

HOT WATER BOILER SCHEDULE							
BOILER NO.	FUEL	BTU/GAL.	BTUH-INPUT	OIL-GPH	IBR-NET	IBR-GROSS	FLUE
*A-504	N°2 OIL	261 MBH	300,150	2.55	227,000	261,000	8"x 12"

* AMERICAN STANDARD

which is also identified in the same manner on the floor plans. The schedule describes the radiator in detail, while the floor plan gives the symbol R/1 only.

• The "Serving" column (the second from the left) indicates that radiator number 1 serves the note department, radiator number 2 serves the vault and office area, and so forth.

• The "Capacity MBH" column indicates the heat output, in British thermal units, of each unit. Radiator number 1 (R/1) has a capacity of 18.1 MBH, or an output of 18,000 Btu/h.

• The "Enclosure Length" column in the schedule gives the physical length of each radiator listed. This information is provided for two reasons: first, to indicate the required length in case the contractor submits a substitute radiator; second, to aid the workmen

on the job to "rough in" for the various pieces of equipment—installing pipes, recesses, etc.

• "Active Length" in this schedule designates the actual length of the radiation tubing inside the housing. It is given in case a substitution is made. In radiator number 1, which has an enclosure length of 24 feet and an active length of 23 feet, 6 inches of space (1 foot total) are on each end of the radiation tubing for pipe connections.

• The column designated "Rating—Btu Lineal Foot" is provided in case another brand of radiator is substituted. The heat output per lineal foot will be kept the same as indicated. Because radiator number 1 has an active length of 23 feet at 790 Btu/foot, the total heat output is 23 × 790 = 18,170 Btu.

• The "Prototype" column gives the

Table 3-11. Many Components and Pieces of Equipment May Be Adequately Described in Schedules Like the One Shown.

			RADIATION		SCHEDULE				
SYMBOL	SERVING	CAPACITY MBH	ENCLOSURE LENGTH	ACTIVE LENGTH	RATING - BTU LIN. FT.	PROTO-TYPE	TYPE OF ENCLOSURE BACK	TYPE OF SLEEVE	REMARKS
R/1	NOTE DEPT.	18.1	24'	23'	790	NESBITT ARCH SILL LINE - STYLE GN12	UNFINISHED	INTERNAL SLEEVE AT MULLIONS	
R/2	VAULT OFFICE	11.8	15-1/2'	15'	790	NESBITT ARCH SILL LINE - STYLE GN12	FINISHED	INTERNAL SLEEVE AT MULLIONS	
R/3	LOAN DEPT.	24.5	32'	31'	790	NESBITT ARCH SILL LINE - STYLE GN12	FINISHED	INTERNAL SLEEVE AT MULLIONS	
R/4	OFFICE	6.5	15'	7'	975	NESBITT ARCH SILL LINE - STYLE FBN-14	UNFINISHED	NONE	ENCLOSURE FULL LENGTH OF WALL
R/5	TELLERS	27.6	36'	35'	790	NESBITT ARCH SILL LINE - STYLE GN12	FINISHED	INTERNAL SLEEVE AT MULLIONS	
R/6	TOILET	2.1	3'	3'	780	NESBITT ARCH SILL LINE - STYLE G-10A	UNFINISHED	NONE	

Table 3-12. Cabinet Unit Heater Schedule.

CABINET UNIT HEATER SCHEDULE

No	SERVING	CFM	CAPACITY MBH	GPM	FAN MOTOR WATTS	FAN MOTOR V/∅/H₂	TYPE	PROTOTYPE	REMARKS
UH 1	BASEMENT REAR ENTRANCE	200	15.0	1.0	110	120/1/60	RECESSED HORIZONTAL CLG. MTD.	TRANE MODEL E46 FORCE FLOW	
UH 2	REAR STAIRWELL	300	20.0	1.0	160	120/1/60	VERTICAL CABINET	TRANE MODEL B42 FORCE FLOW	BUILT-IN THERMOSTAT
UH 3	BASEMENT STORAGE	200	15.0	1.0	110	120/1/60	RECESS HOR CLG. MTD W/O.A CONN	TRANE MODEL E46 FORCE FLOW	PROVIDE BRICK VENT FOR O.A.
UH 4	BASEMENT VAULT	200	15.0	1.0	110	120/1/60	HORIZONTAL EXPOSED CAB CLG. MTD.	TRANE MODEL D16 FORCE FLOW	
UH 5	BASEMENT ELEVATOR LOBBY	200	15.0	1.0	110	120/1/60	VERTICAL CABINET	TRANE MODEL B42 FORCE FLOW	BUILT-IN THERMOSTAT
UH 6	MAIN STAIRWELL	400	30.0	1.5	150	120/1/60	VERTICAL CABINET	TRANE MODEL B42 FORCE FLOW	BUILT-IN THERMOSTAT
UH 7	MAIN VAULT	200	15.0	1.0	110	120/1/60	HOR EXP. CABINET CLG. MTD	TRANE MODEL D16 FORCE FLOW	
UH 8	ENTRANCE	400	30.0	1.5	150	120/1/60	RECESSED HORIZONTAL CLG. MTD	TRANE MODEL E46 FORCE FLOW	
UH 9	BOOK VAULT	200	15.0	1.0	110	120/1/60	HOR. EXP. CABINET CLG. MTD	TRANE MODEL D16 FORCE FLOW	

manufacturer's description of the unit used in the engineer's design. Another manufacturer's unit will usually be allowed, provided it is equal in capabilities to the one originally specified.

• The "Type of Enclosure Back" column will appear only on certain jobs. The reason is that the units specified with an unfinished back will be placed against a solid wall where the back cannot be seen. The finished backs will be installed at the bottom of a floor-to-ceiling glass wall where the backs of the unit can be seen from the outside of the building.

• The "Type of Sleeve" column indicates the blank (nonactivated) connections between the radiator enclosures. Such sleeves are either inside or outside telescoping as needed. They are adjustable up to 7 inches at each sleeve.

• The "Remarks" column is used for added remarks necessary to clearly indicate what is required of the contractor and his workmen. Note that radiator number 4 shall have an enclosure the full length of the wall on which it is to be installed. In some cases a "blank" extension may be necessary to acquire the desired length.

CABINET UNIT HEATER SCHEDULE

The schedule shown in Table 3-12 was used on a set of architectural drawings to describe the nine different unit heaters installed on the project. Each of the unit heaters described in this schedule contained a hot-water coil and a fan to circulate the air over the coil. The hot-water coil received its heated water (180 degrees Fahrenheit) from a centrally located hot-water boiler similar to the one described in Table 3-10.

Chapter 4

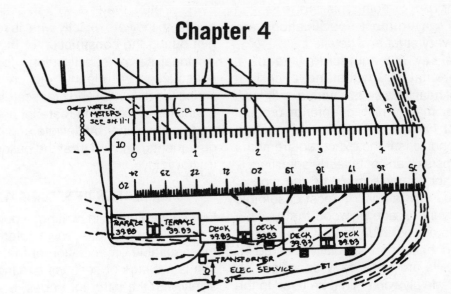

Specifications and Notations

WRITTEN SPECIFICATIONS THAT ACCOMPANY some drawings are the descriptions of work involved, types and grades of materials, and other pertinent information to better help the project to be completed as the designer intended. Written specifications for a building construction project are the descriptions of work and duties required of the architect, engineer, or owner. Along with the working drawings of the project, these specifications form the basis of the contract requirements.

Specifications accompany nearly all drawings for government projects, as well as those for most manufacturers, to ensure that the completed object, product, or service will be exactly as the buyer wants. Let's take a certain type of ammunition. The U.S. government may issue a drawing of a particular cartridge giving all dimensions, caliber, etc.

This cartridge could be made out of inferior materials and cause problems. If bids are solicited, written specifications will accompany the drawing and specify the grade of brass used in the case; the type of primer used; the weight, shape, and type of bullet; and the amount and type of powder, along with the muzzle velocities. The specifications may state that a certain range of the loaded cartridges will be guaranteed along with certain penetration characteristics of the bullet. In most cases the government will want the cases made from virgin brass. If this requirement is not specified, a contractor may use scrap brass in manufacturing the cases that contains more impurities and, generally, is an inferior type of brass. Cartridges loaded to maximum pressures, say, 55,000 pounds per square inch, could cause the inferior cases to rupture—damaging the shooter or rifle, or both. This is one reason why U.S. government specifications are normally very strict and detailed.

Let's say a manufacturer needs a transistor that will withstand a certain amount of heat for a certain period of time. A low bidder may furnish an inferior product that might not hold up under the required design conditions if the exact requirements of the transistor are not clearly specified.

Another example might be the type of plywood used in a boat. In most cases this will be Marine Grade plywood that uses a special waterproof glue to joint the panels. The wood has been specially treated. If these details are not clearly specified, an inferior grade plywood might be used in the boat. It would probably separate in very short order, making the boat unsafe and worthless.

Specifications for building construction should give information regarding the scope of the work to be done by the contractor and the method of carrying out the work, both as to workmanship and material. They should also contain all information (in addition to that shown in the working drawings) necessary for the builder to quickly and accurately estimate the cost of all labor and material for the project. Further information will usually include such requirements as building permits, liabilities for loss, and other items of similar nature. The specifications should also give information regarding the quality of the materials to be furnished and the size of some materials not given on the plans. Every detail of construction—especially methods used—is seldom necessary or included.

Although many people will normally need to see the specifications for any building project (inspectors, material suppliers, owners, etc.), they are written mainly for the builder to indicate exactly what is required of him during the construction of the building. An ideal set of specifications for building construction will clearly and precisely describe the essential and practical limits of the quantities of work and material needed, will protect the owner by definite and proper requirements, and will treat the builder or contractor fairly.

ORGANIZATION OF SPECIFICATIONS

For convenience in writing, speed in estimating work, and ease in reference, the most suitable organization of the specification is a series of sections dealing successively with the different trades, and in each section grouping all the work of the particular trade to which the section is devoted. All the work of each trade should be incorporated into the section devoted to that trade. Those people who use the specifications must be able to find all information needed without taking too much time in looking for it.

The Construction Specification Institute (CSI) developed the Uniform Construction Index some years ago that allowed all specifications, product information, and cost data to be arranged into a uniform system. This format is now followed on most large construction projects in North America. All construction is divided into 16 sections, and each section has several subsections. The following outline describes the various sections normally included in a set of specifications for building construction.

Division 1—General Requirements. This division summarizes the work, alternatives, project meetings, submissions, quality control, temporary facilities and controls, products, and the project closeout. Every responsible person involved with the project should become familiar with this division.

Division 2—Site Work. This division outlines work involving such items as paving, sidewalks, outside utility lines (electrical, plumbing, gas, telephone, etc.), landscaping, grading, and other items pertaining to the outside of the building.

Division 3—Concrete. This division covers work involving footings, concrete formwork, expansion and contraction joints, cast-in-place concrete, specially finished concrete, precast concrete, concrete slabs, and the like.

Division 4—Masonry. This division covers concrete, mortar, stone, masonry accessories, and the like.

Division 5—Metals. Metal roofs structural metal framing, metal joists, metal decking, ornamental metal, and expansion control normally fall under this division.

Division 6—Carpentry. Items falling under this division include: rough carpentry, heavy timber construction, trestles, prefabricated structural wood, finish carpentry, wood treatment, architectural woodwork, and the like. Plastic fabrications may also be included in this division of the specifications.

Division 7—Thermal and Moisture Protection. Waterproofing is the main topic discussed under this division. Other related items such as dampproofing, building insulation, shingles and roofing tiles, preformed roofing and siding, membrane roofing, sheet metal work, wall flashing, roof accessories, and sealants are also included.

Division 8—Doors and Windows. All types of doors and frames are included under this division: metal, plastic, wood, etc. Windows and framing are also included along with hardware and other window and door accessories.

Division 9—Finishes. Included in this division are the types, quality, and workmanship of lath and plaster, gypsum wallboard, tile, terrazzo, acoustical treatment, ceiling suspension systems, wood flooring, floor treatment, special coatings, painting, and wallcovering.

Division 10—Specialties. Specialty items such as chalkboards and tackboards; compartments and cubicles, louvers and vents that are not connected with the heating, ventilating, and air conditioning system; wall and corner guards; access flooring; specialty modules; pest control; fireplaces; flagpoles; identifying devices; lockers; protective covers; postal specialties; partitions; scales; storage shelving; wardrobe specialties; and the like are covered in this division of the specifications.

Division 11—Equipment. The equipment included in this division could include central vacuum cleaning systems, bank vaults, darkrooms, food service, vending machines, laundry equipment, and many similar items.

Division 12—Furnishing. Items such as cabinets and storage, fabrics, furniture, rugs and mats, seating, and other similar furnishing accessories are included under this division.

Division 13—Special Construction. Such items as air-supported structures, incinerators, and other special items will fall under this division.

Division 14—Conveying Systems. This division covers conveying apparatus such as dumbwaiters, elevators, hoists and cranes, lifts, material-handling systems, turntables, moving stairs and walks, pneumatic tube systems, and powered scaffolding.

Division 15—Mechanical. This division includes plumbing, heating, ventilating, and air conditioning and related work. Electric heat is sometimes covered under Division

16, especially if individual baseboard heating units are used in each room or area of the building.

Division 16—Electrical. This division covers all electrical requirements for the building including lighting, power, alarm and communication systems, special electrical systems, and related electrical equipment.

INTERPRETING
CONSTRUCTION SPECIFICATIONS

When reading a set of construction specifications, remember that they are designed to give the grade of materials to be used on a given project and the manner in which they are to be installed or used. Most beginners have a hard time understanding technical terms used in written specifications and also the abbreviated language often used. By utilizing the Glossary for any terms not understood, you should be able to get through a set of specifications with a reasonably good understanding of their contents. As you gain experience, the reading will be easier and your understanding will be better.

Many types and grades of materials listed in construction specifications are by manufacturer's name and catalog or other identifying number. When an item is specified this way, usually an item that is considered an approved equal will be accepted.

INDEX TO SPECIFICATIONS

Construction specifications can vary in size from only a page or two to volumes containing hundreds of pages. To familiarize yourself with all types, visit a local architect or builder and ask to see a set of specifications for a project. If you tell the architect that you are trying to learn to read specifications, he or she may loan you a complete set for study. You should review the following index to a set of construction specifications used on an actual job. This will give you an overview of what exists in construction specifications and prepare you for more serious study with a complete set.

This index is for convenience only. Its accuracy and completeness are not guaranteed, and it is not to be considered as part of the specifications. In case of discrepancy between the index and the specifications, the specifications shall govern.

M-6	Mechanical Plan
M-7	Mechanical Notes and Details
E-7	Electrical Plan and Details
E-8	Electrical Plan, Schedules and Details

SAMPLE SPECIFICATIONS

The following set of construction specifications has been condensed. The specifications will give you a basic knowledge of how such documents appear on actual projects. You are not expected to memorize every word, but by reading them completely through, you will have a general idea of the format and what the specifications contain.

DIVISION 1—GENERAL REQUIREMENTS

Section 1A—Summary of Work

1. GENERAL—The work on this project shall be performed under a single contract. The general contractor shall visit the site and familiarize himself with existing conditions. No extras will be allowed for site conditions that are existing.

2. DESCRIPTION OF WORK—The project generally comprises the construction of a one-story warehouse and office with open shed, covering approximately 24,400 square feet. All rough grading work shown on sheet L-2 of the drawings, including concrete head walls and pipe under access road, will be done by others and is not included in this contract.

3. SHOP DRAWINGS—Submit five copies of shop drawings, unless otherwise noted, of items used on this project to the architect for approval. Submit drawings in accordance with the requirements described in "Bidding Documents." Obtain approval of drawings prior to proceeding with fabrication. These drawings shall indicate the materials, finishes, arrangement, profiles, thicknesses, size of parts, size of openings, assembly and erection details, fastenings, supports, anchors, reinforcements, clearances, hardware coordination, and all necessary connections to work of other trades.

4. SUBSTITUTIONS—The materials or products specified herein and indicated on the drawings by trade name, manufacturer's name, or catalog number shall be provided as specified. Substitutions will not be permitted except as described in "Bidding Documents."

DIVISION 2—SITE WORK

Section 2A—General

1. TEST BORING:

   a. Subsurface soil investigations have been made at the project site. Locations and information obtained are included at the end of this section.

   b. Soils information was obtained for use in preparing foundation design only and does not form a part of the contract documents.

2. BENCH MARK—Contractor shall provide a bench mark in accordance with general note five on drawing L-3.

3. LAYING OUT:

   a. Accurately locate the building on the site according to information given in documents. Erect substantial batter boards, and set stakes securely, to remain in place until building corners and heights are permanently established.

   b. Allocate storage and working areas to avoid interference with subsequent operations. Arrange materials storage in approximate order of use to avoid excessive rehandling.

DIVISION 3—CONCRETE

Section 3A—Formwork

1: MATERIALS:

    a. Exposed concrete surfaces: ⅝ inch or ¾ inch thick, exterior grade, medium or high density resin fiber overlaid Douglas fir plywood.

    b. Concealed concrete surfaces:

        (1) Plywood: DFPA-EXT "plyform," Grade BB, ⅝ inch or ¾ inch thick, as manufactured for formwork.

        (2) Boards: "T and G," Commercial-Standard Douglas fir or number 2 common or better lumber uniform nominal 1-inch thickness, to provide tight forms.

    c. Use adjustable form ties with working strength of not less than 3000 psi. Metal not permitted closer than 3.4 inches to finished surfaces. Do not use ties or spreaders that will leave a hole larger than ⅞ inch in exposed surfaces. Wire ties not permitted.

Section 3B—Concrete Reinforcement

1. STANDARDS—Meet requirements and recommendations of applicable portions of standards listed.

    1. American Concrete Institute (ACI).

    2. Concrete Reinforcing Steel Institute (CRSI).

    3. American Society for Testing and Materials (ASTM).

DIVISION 4—MASONRY

Section 4A—Mortar

1. Types of Mortar—Mortar shall be mixed in the proportions of 1 part portland cement, ¼ part hydrated lime, and between 2-¼ and 3 parts sand by volume.

2. Lay block and brick plumb, level, true to a line, in common bond, or as indicated. Align on exposed face or as indicated. Exposed cells, cracked block, and broken block are not permitted in work that will remain exposed.

3. Finish flush, block joints that will not remain exposed. Rub joints with burlap, where blockwork will have additional finish, to make joints match the texture of the block. Use concave tooled joints for masonry work that will remain exposed. Tool vertical joints first. Joints shall be of a width to suit the coursing with vertical joints of the same size as horizontal.

DIVISION 5—METALS

Section 5A—Structural Metal

1. STANDARDS:

    a. Meet requirements and recommendations or applicable portions of standards listed.

        (1) American Institute of Steel Construction (AISC).

        (2) American Society for Testing and Materials (ASTM).

        (3) American Welding Society (AWS).

2. MATERIALS:

    a. All structural steel columns, beams, angle and beam lintels, and plates shall conform to "Structural Steel, ASTM A36" specifications.

    b. Steel pipe columns shall conform to ASTM A53 specifications, types E or S, Grade B, open-hearth or basic oxygen steel with sulfur not to exceed 0.05 percent. See drawings for location and sizes.

3. FABRICATION—It shall be in accordance with standard specifications of the American Institute of Steel Construction according to the latest edition.

DIVISION 6—CARPENTRY

Section 6A—Rough Carpentry

1. MATERIALS:

a. Framing Lumber—Grade marked, SPIB, number 2 common Southern yellow pine, moisture content under 19 percent.

b. Bracing and Blocking—Grade marked, SPIB, number 3 common Southern yellow pine moisture content under 19 percent.

c. Plywood—Fir plywood graded INT-DFPA-A exposed faces and B concealed faces for interior use and EXT-DFPA for exterior use. Plywood for paint finish shall be EXT-DFPA-MD overlay.

2. PRESERVATIVE:

a. Treat wood that will be in contact with masonry, steel, or concrete, and treat wood used in roof and cornice construction with Doppers Company Wolman Salts with retention of 0.35 pounds per cubic foot dry salt. Brush coat surfaces that have been cut after treatment.

3. ROUGH HARDWARE—Provide all rough hardware items required to permanently secure all rough carpentry. Items shall include all types and sizes of nails, screws, nuts and bolts, washers, anchors, and other items.

Section 6B—Finish Carpentry

1. MATERIALS:

a. Wood Trim—It shall be "choice" Idaho white pine or "C" select ponderosa pine.

b. Plywood—For general interior use it shall be Douglas fir plywood, interior B-B faces where both sides are exposed and B-D faces where one side exposed.

c. Hard Wood Trim—Where hardwood trim, edges and/or facing are indicated, it shall be first grade maple or birch.

d. Millwork:

(1) Shelving—It shall be ¾-inch custom grade white pine free from knots, sap, or other defects, and sanded and ready for the painter's finish.

(2) Cabinetwork:

(a) Lumber: custom grade white pine.

(b) Douglas fir plywood: custom grade.

(c) Hardwood plywood: economy grade.

(3) Door Frames—These shall be ¾-inch thick custom grade white Pine with applied white pine stops. Omit stops at all cased openings.

2. WORKMANSHIP:

a. General:

(1) Provide workmanship complying with Architectural Woodwork Institute "premium" grade for "natural" finish, "custom" grade for "painted" finish.

(2) Make joints neatly and carefully with surfaces straight and clean. Sand work with the grain and remove machine marks by sanding. Eliminate cross scratches.

(3) Glue shop-assembled surfaces and glue-block at concealed locations.

(4) Make joints and connections by best approved practices of the cabinetmaking trade, including dadoes and mortises and tenons where possible.

Section 6C—Custom Woodwork

1. CUSTOM WOODWORK:
a. General Provisions:

(1) Fabric cabinets, vanities, and bookcases as detailed from sound, kiln-dried lumber and plywood of species and grades specified.

(2) Backs shall be ¼-inch thick plywood unless specified otherwise.

(3) Stile and rail construction shall be solid wood of species specified.

(4) All references to grade and species shall conform to AWI standards for species selected.

b. Shelving and Doors in Office Project Room:

(1) Construct according to details of species of sound material in grades as follows:

(a) Lumber: custom grade white pine.

(b) Douglas fir plywood: custom grade.

(c) Hardwood plywood: economy grade.

(2) Hardware and Accessories:

(a) Adjustable shelf standards: K and V number 255, nickle-plated, flush installation.

(b) Adjustable shelf supports: K and V number 256, nickel-plated.

(c) Magnetic catches for projection room windows: Stanley Hardware number 46ALD.

(d) Hinges for projection room windows: minimum two each door.

2. CUSTOM PANELING:
a. Material--It shall be Weldwood Algoma grade, plain sliced, walnut, ¼-inch 3 ply, 4 foot wide by necessary lengths to cover floor to ceiling, mineral core construction, fire-retardant panels (Class I) as manufactured by U.S. Plywood. All panels shall be good on one side, bonded with heat-resistant glue, and lot matched with no sapwood allowed and numbered. All paneling shall be surfaced with clear Permaguard. Each panel shall bear Underwriters' label for Class I.

b. Location—See finish schedules for location of paneling.

c. Protection—The contractor shall take special care so that the paneling is not damaged by operations subsequent to its installation.

DIVISION 7—MOISTURE CONTROL

Section 7A—Dampproofing

1. MATERIAL—It shall be "Moistop" as manufactured by the Sisalkraft Division of St. Regis Paper Company, conforming with ASTM C 171.

2. PLACING—"Moistop" shall be applied over level and tamped stone or earth base under all interior floor slabs and/or as detailed. It will be applied with the width of the roll parallel to the direction of the pour of concrete. All joints should be lapped 6 inches, sealed, and turned up 3 inches at walls.

3. PROTECTION—Contractor shall exercise extreme caution to prevent displacement or puncture.

Section 7B—Building Insulation

1. MATERIALS:

a. Blanket Insulation: roll or batt type, glass or mineral wool composition with vapor barrier, fire-resistant; 4 inches nominal thickness unless otherwise noted.

b. Perimeter Insulation: inorganic glass fiber board type, 24 inch × 48 inch (or

60 inch) × 1 inch thick, scored, 8-10 pound density; federal specification HH-1-562, Type 1, Class 2 as amended.

c. Insulated Liner—⅞-inch thick rigid mineral insulation applied to 26-gauge, zinc-coated embossed-painted steel as manufactured by Armco Steel and all accessories necessary to complete the installation in accordance with the manufacturer's guide specification number 17.

2. INSTALLATION:

a. Blanket Insulation:

(1) Place vapor barrier toward interior (warm side).

(2) Install carefully, in full thickness, with all joints lapping or butted tightly and blankets tight to all adjacent surfaces.

(3) Wire, clip, staple, or adhesively apply to metal lath and other construction where necessary to properly position and secure blankets.

Section 7C—Preformed Siding

1. SCOPE—It includes all metal siding (except that in pre-engineered buildings) as well as rooftop mechanical equipment fences.

2. MATERIALS:

a. Metal Siding (everywhere except office overhanging roof facing)—It shall be 3 inches deep by 16 inches wide number 362S sculptured Steelox panels as manufactured by Armco Steel.

b. Metal Siding (office overhanging roof)—It shall be 3 inches deep by 12 inches wide number 322 Steelox panels (ribs out) as manufactured by Armco Steel.

c. All accessories necessary to installation of siding, as manufactured by Armco Steel.

3. INSTALLATION—It shall be in accor-

dance with details and recommendations of the manufacturer, including all accessories necessary to complete the installation as specified.

a. Zee Base—It shall be fastened to the top of concrete or masonry with ⅜-inch diameter bolts at 48 inches on center.

b. Roof Angle—Siding shall be fastened to an angle at 16 inches on center with rib clips.

Section 7D—Membrane Roofing

1. EXAMINING:

a. Examine surfaces that are to receive insulation and roofing. Report unsatisfactory conditions.

b. Do not start roofing and insulating work until unsatisfactory conditions have been corrected.

2. DELIVERING AND STORING:

a. Deliver packaged materials to the site in manufacturer's original, unopened, labeled bundles or containers.

b. Arrange deliveries to provide sufficient quantities to permit continuity of any phase of work.

c. Store to prevent damage to materials or structure. Do not store materials on roof construction in concentrations large enough to impose excessive stress on decking or structural members.

Section 7E—Sheel Metal Work

1. MATERIALS:

a. Galvanized Steel—ASTM A 361 treated to hold paint, not less than 26 gauge.

b. Copper—ASTM B 152 weighing not less than 16 ounces per square foot.

c. Lead-Coated Copper—As above, coated both sides with lead weighing not less than .06 pounds per square foot.

2. METAL ROOF FLASHING:

a. Metal Cap (Counter) Flashing—It shall be two-piece, 26-gauge galvanized metal at the intersection of any roof and vertical masonry surface. Cap flashing for roof ventilators shall be an integral part of the ventilator as furnished by others. Flashing shall be formed in sheets a maximum of 8 feet long. Lap end joints 2 inches and solder. Flashing at angles shall be continuous. Cap flashing shall overlap base to the junction of cant and vertical wall.

b. All vents and pipes through the roof shall be flashed using lead-coated copper tube with an 18-inch square flange soldered to the tube.

c. Install flashing as indicated and as necessary to obtain a weathertight condition. Install items specified in other sections as furnished for installation with sheet metal work.

d. Fabricate sections up to 8 feet in one piece. For sections over 8 feet, use sheets as large as are practical. Fold exposed edges back, not less than ⅜ inch, and flatten.

e. Use minimum 26-gauge galvanized steel for metal flashing and counterflashing.

Section 7F—Caulking and Sealants

1. EXAMINING:

a. Examine surfaces that are to be caulked. Report unsatisfactory conditions.

b. Do not start caulking until unsatisfactory conditions have been corrected.

2. DELIVERING AND STORING:

a. Deliver packaged materials to the site in manufacturer's original, unopened, labeled containers.

b. Arrange deliveries to provide sufficient quantities to permit continuity of any phase of work.

c. Store and handle caulking items to prevent damage to materials or work in place.

3. PREPARING:

a. Properly prepare joints and surfaces to receive caulking compound. Mask or protect as necessary to prevent smearing of adjacent surfaces.

b. Remove dust, moisture, rust, grease, glaze, and loose materials that could interfere with adhesion of compound.

c. Rake joints as necessary to obtain a minimum of ¼ inch for caulking. Maintain caulking width of not more than 1 inch.

DIVISION 8—DOORS, WINDOWS AND GLASS

Section 8A—Metal Doors and Frames

1. EXAMINING:

a. Examine surfaces and openings that are to receive metal doors or frames. Report unsatisfactory conditions.

b. Do not start installation of doors or frames until unsatisfactory conditions have been corrected.

2. DELIVERING AND STORING:

a. Deliver packaged materials to the site in manufacturer's original, unopened, labeled containers.

b. Store metal doors in a vertical position with not less than ¼-inch space between units. Store metal doors to maintain dry, ventilated condition.

c. Store to prevent damage to materials or structure.

3. HOLLOW METAL DOORS AND FRAMES:

a. Frames—For masonry walls there shall be double rabbeted standard F-16 and F-14 flush Steelcraft frames of sizes shown on the drawings with 4-inch head. Frames for wood stud partitions shall be double rabbeted frames Steelcraft DW-16 of sizes shown on the drawings. Masonry frames shall be constructed of 14 or 16-gauge cold rolled steel, mitered, and be reinforced at corners with a snap lock corner clip. Mortise for hinges and strikes. Provide 10-gauge hard tempered steel reinforcements and strike cutouts. Jambs should come flush to the floor with anchors. Provide three adjustable "T" wall anchors per jamb, shipped loose, for masonry openings. Predrill interior jamb for three silencers per jamb to be provided with hardware.

b. Doors—These shall be series LF-18-4 manufactured by Steelcraft of seamless construction, 1-¾ inch thick, full flush of sizes shown on the drawings. The doors shall be constructed of 18-gauge cold rolled steel sheet with an impregnated kraft honeycomb core and 14-gauge channels top and bottom. Reinforce for hardware as specified. Outswinging exterior doors shall have top caps.

Section 8B—Wood Doors

1. PROTECTING:
a. Handle doors, sash, and frames to avoid injury to persons and damage to materials or work in place. Satisfactorily repair or remove and replace work that has been damaged.
b. Protect finished surfaces and edges of doors.
2. FLUSH WOOD DOORS:
a. All flush wood doors shall be 7-ply Weldwood Novodors as manufactured U.S. Plywood, a division of U.S. Plywood Champion Papers, Inc.
b. Faces—Hardwood face veneers shall be walnut to match Weldwood walnut flush paneling, thoroughly dried and laid with grain at right angles to grain of the cross

bands. Cross bands shall be thoroughly dried hardwood 1/10 inch thick extending the full width of the door and laid with the grain at right angles to the face veneers.
c. Core—It shall be a single panel thickness of 3-ply particle board. Lineal expansion under ASTM Test D 1037, Section 76-79, shall not exceed 0.20 percent in either direction. Modulus of rupture shall be a minimum of 900 PSI in either direction. Faces shall be of 0.010-inch thick flakes and with resin content of 14 percent with minimum density of 47 pounds/cubic foot. Inner particle core shall have minimum resin content of 7 percent maximum density of 25 pounds/cubic foot.

Section 8C—Special Doors

1. COILING DOORS:
a. Doors shall be as manufactured by J.G. Wilson Corporation of types as noted below and sizes as shown on the drawings.
b. Rolling Service Doors—nonlabeled, labeled, and weather doors—shall comply with J.G. Wilson specifications for the various types, as noted in their current catalog.
c. Location:
(1) Door number 7 shall be model number BJA-1006-AM-WD, under lintel, flat slat weather door, chain gear operated with dust-retarding rubber flaps at guides, hood baffle, and bottom bar astragal.
(2) Door number 29 shall be model number BJA-1002-AA-FS, under lintel, nonlabeled rolling service door with nonweathering flat slat. It shall have a bottom bar astragal.
(3) Door number 69 shall be model number A-1707-FS crank-operated, "B"-labeled, face-mounted rolling service door with fusible link. The

manufacturer is to provide an over-size certificate of inspection.

(4) Doors numbers 70 and 71 shall be model number A-1704-FS, manual-operated, "B"-labeled, face-mounted, rolling service doors with fusible link.

2. OVERHEAD DOORS:

a. Manufacture—Doors shall be nominal 2 inches thick, aluminum frame, fiberglass doors as manufactured by Overhead Door Corporation. Fiberglass Industrial doors shall be as sized on the drawings.

b. Provide standard 2-inch track for all doors under 20 feet wide and 3-inch track for all doors 20 feet or wider. Track shall be high lift type, except where noted as vertical lift it shall be full vertical track. All track assemblies and locations shall be verified with and coordinated by other trades and items of construction.

Section 8D—Finish Hardware

1. SCOPE:

1. The work shall consist of furnishing and delivering all finish hardware required to satisfactorily complete this project in strict accordance with the following specifications. It is not intended that this specification mention each item of hardware required, but it is intended to establish type and quality for the principal locations and type of openings where hardware will be applied. Any item of finish hardware not specifically mentioned but necessary for the proper completion of this project shall be provided at no additional charge to the owner and/or architect.

b. Work Not Included: rough hardware, hardware for windows and toilet compartments, and casework hardware.

2. SERVICE—Upon completion of the job, a qualified representative of the hardware supplier shall visit the job site, make necessary adjustments to the hardware, report any misapplications he may notice to the architect, and review the keying system, proper adjusting procedures (locksets, closers, and exit devices), and maintenance procedures with the owner and custodial staff. He shall also turn over a complete set of adjusting tools to the owner.

3. MATERIALS:

a. Butts—Butts to be furnished are in accordance with the following unless specifically noted otherwise in the outline schedule:

(1) Exterior doors shall have BB1193 $4\frac{1}{2} \times 4\frac{1}{2}$.

(2) Interior doors shall be equipped with closers. All doors 3 feet wide and over shall have BB1279.

(3) Furnish $1\frac{1}{2}$ pair butts per door leaf and 2 pair for dutch doors and doors over 7 feet, 6 inches high.

(4) Pairs of doors are to be furnished with the same type of butts for each leaf.

b. Locksets:

(1) Locksets are to be furnished with screwless sectional trim equal to Sargent's Cleveland (LL) design and LPJ lever handle design in US26D finish. All locksets are to be furnished with curved lip strikes, wrought box strikes, and an antifriction type latch bolt.

(2) All dead bolt function locksets must have an antipanic feature whereby the inside knob retracts the dead bolt and latch bolt simultaneously.

(3) All locksets shall be grand-master-keyed, master-keyed, and keyed alike in sets as later directed

by the architect and owner. All locks are to be factory keyed, and complete records are to be kept on file at the factory for future reference. Furnish eight master keys for each building and four grandmaster keys.

Section 8E—Operators

1. GENERAL—Furnish and install pneumatic operators, controls, and accessories for doors 63 and 64 at plant and office, including necessary connecting hardware and air supply (by others), as indicated on architectural plans and in accordance with manufacturer's working drawings, wiring diagrams, and instructions. Equipment shall be as manufactured or supplied by Stanley Door Operating Equipment, Division of The Stanley Works, New Britain, Connecticut.

2. OPERATION:

a. Operators—Air power opening and spring closing. These shall be pneumatically powered from central air supply. Opening action shall be controlled by built-in pressure regulator set and shall be adjustable for pressure and volume for required speed and power at door location. Closing speed shall be controlled by power springs and valve exhaust adjustment. There are built-in, adjustable two-stage checking cylinders for both opening and closing limits. Also, there are operators to contain plug-in type electrical control relays and operators to instantaneously recycle to full open position from any point in closing cycle. Operator shall be Underwriters' Laboratories listed.

Section 8F—Glass and Glazing

1. MATERIALS:

a. Plate Glass—It shall be polished, glazing quality, ¼-inch thickness as man-

ufactured by Pittsburgh Plate Glass Co. L.O.F. or approved equal.

b. Clear Wire Glass. It shall be polished Misco of ¼-inch thickness as manufactured by Mississippi Glass Co. or approved equal.

2. INSTALLING GLASS:

a. Prime rebates in wood or steel before setting glass. Clean and prepare surfaces to receive putty or compound.

b. Secure glass with points, clips, or beads as indicated. Set glass in bed of putty or compound, so that the putty or compound completely surrounds glass.

Section 8G—Storefront System

1. MATERIALS:

a. Framing members, mullions, transition members, etc., shall be Kawneer Narrow Line as shown on the drawings. Stops, thresholds, doors, and miscellaneous accessories shall be Kawneer products.

2. INSTALLATION:

a. Furnishing and installing glazed frames and entrances shall be by a frame manufacturer or by an approved representative of a frame manufacturer.

b. All items shall be set level, plumb, square, and at proper elevations and in alignment with other work.

DIVISION 9—FINISHES

Section 9A—Lath and Plaster

1. MATERIALS—They shall be United States Gypsum or approved equal.

a. Lath

(1) Cut from copper-bearing steel sheets coated with rust-inhibiting paint after cutting, or cut from zinc-coated steel sheets; 3.4-pound, ⅜-inch rib base for spans up to 16

inches; 4.0-pound rib lath for spans over 16 inches, maximum spans not to exceed 24 inches. Reinforce all corners of openings with metal lath—12 inch × 24 inch placed diagonally.

b. Plaster

(1) Water should be clean, fresh, and suitable for domestic consumption.

Section 9B—Gypsum Drywall

1. INSTALLING:

a. Cut wallboard by scoring and breaking or by sawing. Work from face side. Scribe wallboard to fit abutting surfaces.

b. Install wallboard with the long edge perpendicular to supporting members. Use longest pieces practical Stagger Joints on opposite sides of a wall. Provide support for all edges.

2. JOINTS—Apply joint compound sufficiently thick to hide board surface at internal angles and butt joints. Cover nail heads and depressions with compound.

Section 9C—Tile Work

1. MATERIALS:

a. Basic Materials: by American Olean or equal.

b. Wall Tile: glazed, 4¼ inches by 4¼ inches, standard grade, standard colors as selected.

2. INSTALLING:

a. Set floor tile on slabs on the ground with cement mortar. Comply with TCA installation method 102.

b. Set wall tile over gypsum wallboard with organic adhesives. Comply with TCA installation method 206.

Section 9D—Acoustical Treatment

1. MATERIALS:

a. Ceiling tile shall be 24-inch × 48-inch × ⅝-inch panels. Tile shall meet the requirements of federal specification SS-A-118b and have an Underwriters' Laboratories rating of "Class A Incombustible." The tile shall be Armstrong "Minaboard" or approved equal.

2. INSTALLATION:

a. Ceiling tile shall be installed in 24-inch × 48-inch regular pattern using the .031 web ten ceiling grid.

b. Ceiling tile shall be installed symmetrically in areas shown on the plans.

Section 9E—Resilient Flooring

1. MATERIAL:

a. Floor Tile—Excelon, standard, vinyl-asbestos tiles, ⅛ gauge 9 inch × 9 inch manufactured by Armstrong or approved equal. Colors are to be selected by the architect.

b. Vinyl Base—It shall be ⅛-inch gauge, 4-inch high vinyl cove base with preformed internal and external corner pieces. Base shall be Armstrong, Johnson, or approved equal.

2. INSTALLATION:

a. All work must be done in strict accordance with the manufacturer's specifications. Take special care to use the proper adhesive.

b. Lay out tile symmetrically about centerlines of rooms or spaces. Do not cut tile except at the border, with no tile less than 4 inches wide. Provide a black strip at doors where color changes are made from one room to another.

Section 9F—Painting

1. PREPARATION OF SURFACES:

a. Wood—Sandpaper to a smooth and even surface. Dust off well. After priming has been applied, fill nail and other holes and cracks with putty.

b. Steel and Iron—Remove all grease, scale, and rust. Dust and touch up any chipped or abraded places on items that have been shop-coated.

2. MATERIALS:

a. Materials shall be first quality products as manufactured by Benjamin Moore, Sherwin-Williams, Glidden, Pratt and Lambert, Pittsburgh, Martin Senour, or Dutch Boy. Trade names are used to designate the quality desired only. All materials shall be used directly from original containers without dilution.

Division 10—Specialties

Section 10A—Toilet and Janitor Accessories

1. TOILET COMPARTMENTS:

a. Compartments shall be Weis "Hi-Stile, Weisteel Flush" metal compartments with baked enamel finish and Fabcore construction.

b. Brackets shall be extra heavy aluminum extrusions, etched and anodized.
2. TOILET ACCESSORIES

a. Locations of these accessories shall be verified with the architect prior to installation.

b. Toilet Paper Holder—The holder is adjacent to each water closet, Model B-274 (double roll capacity) by Bobrick Dispensers, Inc.

DIVISION 13—SPECIAL CONSTRUCTION

Section 13A—Pre-engineered structures

1. Scope:

a. These specifications cover the material for the fabrication of steel buildings so designed and constructed as to be weathertight, easily erected, and capable of being dismantled and re-erected.

b. The materials furnished shall include the bracing, roofing, siding, anchor bolts, doors, windows, hardware, fasteners, caulk and, when specified, any other component parts for the metal building only, including the erection of same, if so stated in the proposal. All materials shall be new, unused, and free from defects and imperfections. The materials should be fabricated in a workmanlike manner.

DIVISION 14—CONVEYING SYSTEMS

Section 14A—Lifts

1. SCOPE—Contractor shall be responsible for providing concrete pit, curb angle framing, and concrete work in accordance with details. He shall install all bolts, etc., as provided by the dockboard manufacturer.

DIVISION 15—MECHANICAL

Section 15-1—General Provisions

15.1.1 General and Special Conditions

A. The architectural general and special conditions before submitting his proposal.

B. The general contractor shall be responsible for all work included in this section. The delegation of work to the mechanical contractor shall not relieve him of this responsibility. The mechanical contractor and his subcontractors who perform work under this section shall be responsible to the general contractor.

C. Where items of the general conditions or of the special conditions are repeated in this section of the specifications, it is intended to call particular attention to or qualify them. It is not intended that any other parts of the general conditions or special conditions shall be assumed to be omitted if not repeated herein.

15.1.2 Scope of the Work

A. The scope of the work included under this section of the specifications shall include a complete plumbing, heating, ventilating, and air conditioning system as shown on the plans and as specified herein. The mechanical contractor shall provide all supervision, labor, materials, equipment, machinery, and any other items necessary to complete the system. The mechanical contractor shall note that all items of equipment are specified in the singular; however, the contractor shall provide and install the pieces of equipment as indicated on the drawings and as required for a complete system.

B. It is the intention of the specifications and drawings to call for finished work that has been tested and is ready for operation. Wherever the word "provide" is used, it shall mean "provide and install complete and ready for use."

C. Any apparatus, appliance, material, or work not shown on the drawings but mentioned in the specifications, or vice versa, or any incidental accessories necessary to make the work complete and perfect in all respects and ready for operation, even if not particularly specified, shall be furnished, delivered, and installed by the contractor without additional expense to the owner.

D. Minor details not usually shown or specified, but necessary for proper installation and operations, shall be included in the contractor's estimate, the same as if herein specified or shown.

E. With submission of a bid, the contractor shall give written notice to the architect of any materials or apparatus believed inadequate or unsuitable in violation of laws, ordinances, rules, or regulations of authorities having jurisdiction, along with any necessary items or work omitted. In the absence of such written notice, it is mutually agreed that the contractor has included the cost of all required items in his proposal. He will be responsible for the approved satisfactory functioning of the entire system without extra compensation.

DIVISION 16—ELECTRICAL

Section 16.1—General Provisions

16.1.1 General and Special Conditions

A. The architectural general and special conditions for the construction of this project shall be a part of the electrical specifications. The electrical contractor shall examine the general and special conditions before submitting his proposal.

B. The general contractor shall be responsible for all work included in this section. The delegation of work to the electrical contractor shall not relieve him of this responsibility. The electrical contractor and his subcontractors who perform work under this section shall be responsible to the general contractor.

C. Where items of the general conditions or of the special conditions are repeated in this section of the specifications, it is intended to call particular attention to or qualify them. It is not intended that any other parts of the general conditions or special conditions shall be assumed to be omitted if not repeated herein.

16.1.2 Scope of the Work

A. The scope of the work consists of the furnishing and installing of complete electrical systems—exterior and interior—including miscellaneous systems. The electrical contractor shall provide all supervision, labor, materials, equipment, machinery, and any other items necessary to complete the systems. The electrical contractor shall note that all items of equipment are specified in the singular; however, the contractor shall provide and install the pieces of equipment as indicated on the drawings and as required for complete systems.

B. It is the intention of the specifications and drawings to call for finished work that has been tested and is ready for operation.

C. Any apparatus, appliance, material, or work not shown on drawings but mentioned in the specifications, or vice versa, or any incidental accessories necessary to make the work complete and perfect in all respects and ready for operation, even if not particularly specified, shall be furnished, delivered, and installed by the contractor without additional expense to the owner.

D. Minor details not usually shown or specified, but necessary for proper installation and operation, shall be included in the contractor's estimate, the same as if herein specified or shown.

E. With submission of a bid, the electrical contractor shall give written notice to the architect of any materials or apparatus believed inadequate or unsuitable in violation of laws, ordinances, or rules, along with any necessary items or work omitted. In the absence of such written notice, it is mutually agreed that the contractor has included the cost of all required items in his proposal. He will be responsible for the approved satisfactory functioning of the entire system without extra compensation.

Chapter 5

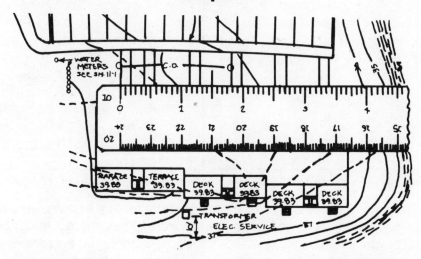

Orthographic Projections

MANY DRAWINGS ARE USED TO CONVEY IN-formation, but orthographic projections are probably the most popular. Although pictorial views of objects give a realistic appearance, only drawings with exact information concerning shape and size and material can be used to properly construct an object. This information is usually best given in a drawing where several related views of an object are presented in the proper manner.

The pictorial drawing of a cube in Fig. 5-1 is a single-view drawing that shows how the object appears to the viewer. One side of the cube is the front and another side is the top. To make an orthographic projection of this cube, you would look first directly at the top, then at the front, and then at the right end.

The top is what you see when you look down directly on the top of the cube. The sides, the bottom, and the ends are cut from your view. Therefore, you draw only what you can actually see from this angle.

The front is what you see when the cube is directly in front of you. If you look directly at the front of the cube, you will not be able to see any other sides but the front side. Try it one time.

If you turn the cube so that you see only the end, you will have a drawing showing only this end. The end of the cube cuts your view from any other sides.

VIEWS

Although Fig. 5-2 shows three separate views in order to demonstrate how an orthographic projection is laid out, an experienced craftsman could construct the cube from wood, steel, or any other material from just a single view. A cube has six equal sides. If you showed one side drawn to scale

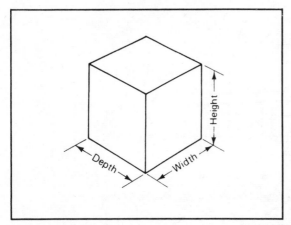

Fig. 5-1. View of a cube as it would appear to the naked eye.

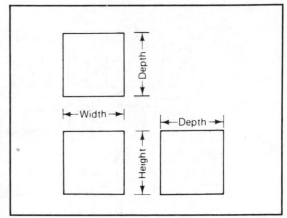

Fig. 5-2. Orthographic drawing of the cube pictured in 5-1.

or else dimensioned, the exact size of the cube would be known from this one drawing.

Let's take the same cube (slightly modified) and cut a section from it; the resulting drawing would appear like Fig. 5-3. Note how this recess in the cube is shown in all views. In the top view observe how the edge of the recess is shown by a solid line, because that is what you see if you look down on the object. The front view shows the out-

line of the recess. The end view shows three horizontal lines representing the edges of the top, bottom, and the recess. All lines are solid because all the edges are seen.

Suppose we further modify our original cube by cutting a slot or keyway in the side representing the front view (Fig. 5-4). This slot is shown in the front view and in the top view by solid lines. The top view shows both the width and the depth of the slot. If you look

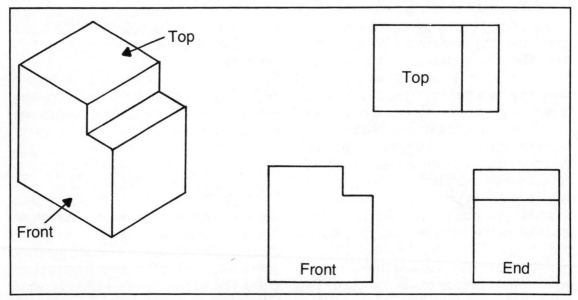

Fig. 5-3. Section cut from the cube shown in Fig. 5-1.

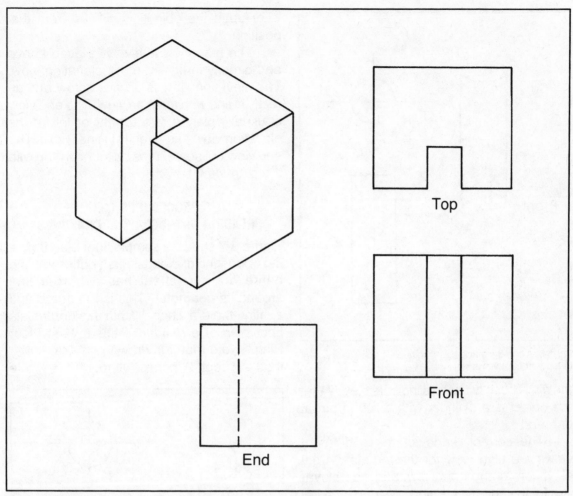

Fig. 5-4. Orthographic view of a cube modified with a keyway inserted.

directly at the front of the object, you will notice that the depth of the slot could not be shown in the front view.

When you observe the end view, note that a dotted line is used to show an invisible edge. It is not possible to see the slot when you look at the end view, but a dotted line is used to indicate that it is there. This dotted line appears at the left of the end view rather than at the right. Its position in the drawing coincides with the position of the slot as shown in the front and top views.

Another example of an orthographic projection is shown in Fig. 5-5. It is a sketch of a cylinder that is cut off square at each end and is shown in two views—top and front. Note that an orthographic projection of the cylinder appears as a circle in the top view and as a rectangle in the front view.

Figure 5-6 is very similar to a cylinder discussed previously, except that it has a pin in the end of it. The length and width of this pin are shown in the front view of the drawing. Note how the two circles in the top view of the drawing represent, first, the outside edge of the cylinder and, second, the edge of

71

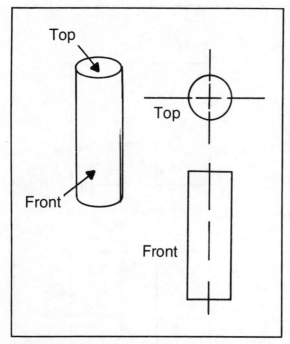

Fig. 5-5. A cylinder that is cut off square at each end is shown in top and front views.

the pin. These two views indicate clearly that the object is a cylinder with a round pin at one end.

The correct terms for these different views are plan view for the top view, front elevation for the front view, and side elevation for the end or side view. We can form the following rule. The plan or top view is the view of the object as you look down on it from above. Any dotted lines in that view indicate hidden edges, corners, or surfaces in or on the object as viewed from that position.

The elevation or front view is the view of the object as you look directly at it from the front. Any dotted lines in that view indicate hidden edges, corners, or surfaces in or upon the object as viewed from that position.

The side elevation or end view is the view of the object when you look at it from the side or end. Any dotted lines in that view indicate hidden edges, corners, or surfaces

in or upon the object as viewed from that position.

The plan view is always placed above and directly in line with the elevation view. The elevation view is always below and directly in line with the plan view. The end view is usually placed to the right or left of the elevation view and directly in line with it. The end view will sometimes be found in line with the plan view.

PROCEDURES FOR READING ORTHOGRAPHIC DRAWINGS

Figure 5-7 shows a sample floor plan from a set of working drawings. First, notice that the entire drawing sheet has a border line around its perimeter. This is to square and confine the drawings. Within this border is a floor plan of a residence titled "First Floor Plan." Note that it is drawn in a scale of ¼ inch = 1 foot, 0 inches (although the whole

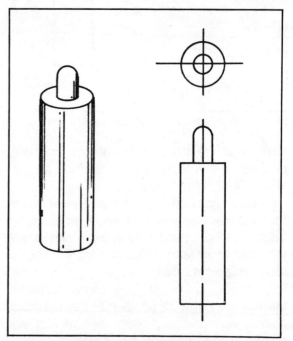

Fig. 5-6. Drawing of a cylinder that has a pin inserted in one end.

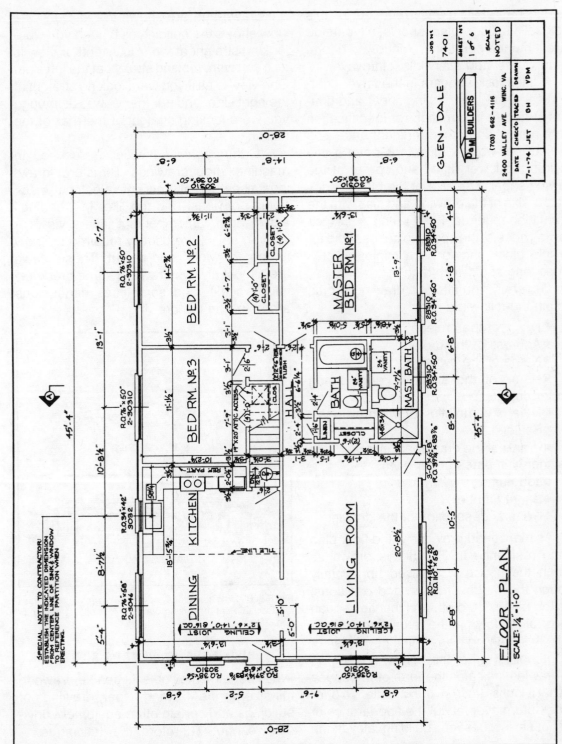

Fig. 5-7. Floor plan of a building.

drawing has been greatly reduced for this book). This floor plan shows all outside walls, interior partitions, windows, doors, toilet fixtures, and stairwells (stairways).

A title block appears in the lower right-hand corner of the drawing sheet and contains sufficient information for identification of the project and other pertinent data. You will find that working drawings have a title block in the lower right-hand corner of the drawing sheet. The size of the block varies with the size of the drawing and also with the information required. For 11-inch × 17-inch drawing sheets, the space to be reserved for the title block should be a minimum of 3 inches long and 1 inch high.

Generally, the title of a drawing should contain the following:

- Name of the project.
- Address of the project.
- Name of the owner or client.
- Name of the architect, engineering firm, or designer.
- Date of completion of the drawings.
- Scale(s).
- Initials of the draftsman, checker, and designer with dates under each.
- Job number.
- Sheet number.
- General description of the drawing.

As mentioned previously, the floor plan of the residence in Fig. 5-7 is drawn as though the house was sliced horizontally through the window, doors, and partitions. The top part is lifted off, and the viewer can look straight down at the remaining portion. While this drawing gives you a good idea of the interior layout of the house, nothing on the drawing indicates the form of the house or what it looks like from the outside. To offer further information about the appearance of the house in question, building elevations are drawn and are shown in Fig. 5-8. One view shows the building as though you were looking straight at the front. Another view is as if you were looking straight at the left side. One view is as if you were looking straight at the right side, and another view is as though you were looking straight at the rear of the building.

Orthographic projections are used in machine shop drawings. Structural drawings, sheet metal drawings, electrical/electronic drawings, and the like. In "reading" any of them, remember that several views of a given object are usually required to show its length, width, and height. These views are almost always drawn to some predetermined scale, and sometimes dimensions are noted on the drawings.

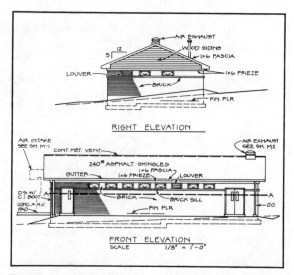

Fig. 5-8. Building elevations are views that are almost always included in a set of working drawings for building construction.

USING THE ARCHITECT'S SCALE

While certain small objects can be drawn to full size on the drawing paper, drawings of buildings and certain other equipment obviously cannot. Therefore, the drawing is reduced in size so that all the distances on the drawing are drawn smaller than the actual dimensions of the building or piece of equipment. All dimensions are reduced in the same proportion. The ratio or relation between the size of the drawing and the size of the object being drawn is indicated on the drawing paper (⅛ inch = 1 foot, 0 inches, for example). The dimensions are the actual dimensions of the building or other object— not the distance that is measured on the drawing.

The most common method of reducing all the dimensions in the same proportion is to select a certain distance. Let that distance represent 1 foot. This distance is then divided into 12 equal parts, each of which represents 1 inch. If ½-inch divisions are required for the drawing (although they rarely are on most electrical drawings for building construction), these twelfths are further subdivided into halves. This method results in a usable scale that represents the common foot rule with its subdivisions into inches. When a measurement is done on the drawing, it is made with the reduced foot rule known as the architect's scale. When a measurement is taken on the building or object itself, it is made with the standard foot rule.

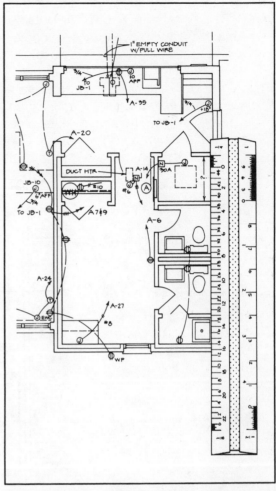

Fig. 5-9. Drawings are read by placing the appropriate scale on the drawing and reading the figures.

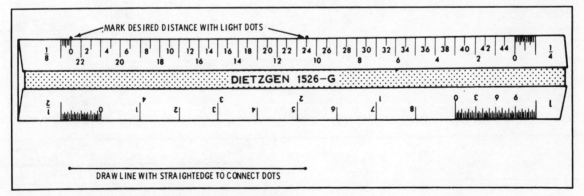

Fig. 5-10. An architect's scale showing four different degrees.

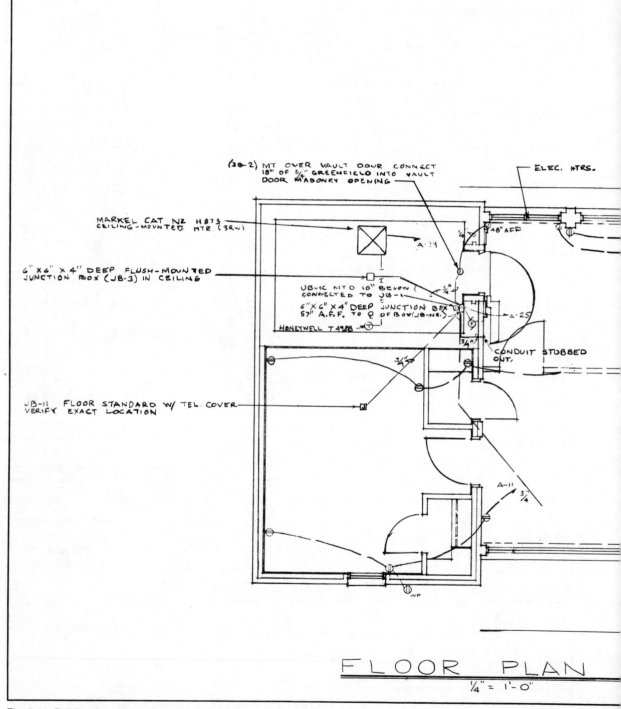

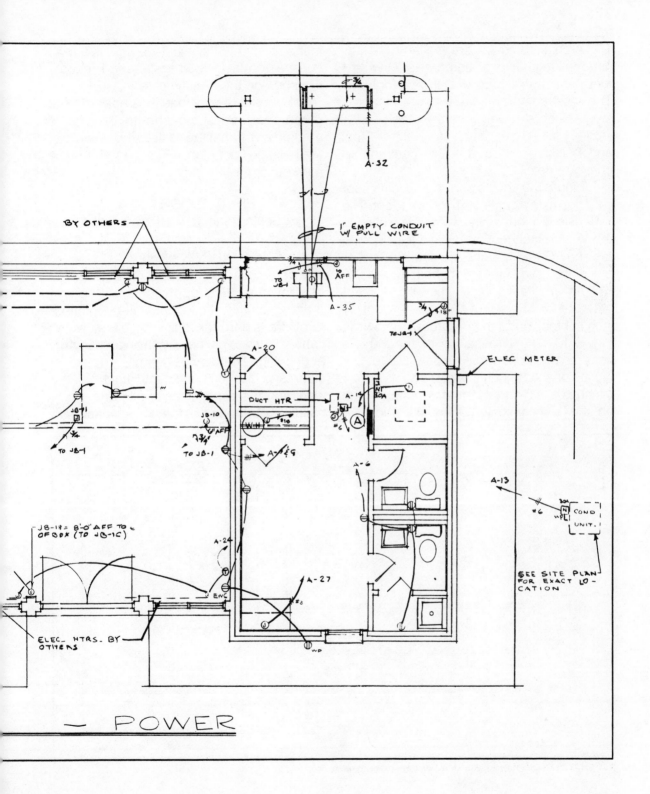

BY OTHERS

A-32

1" EMPTY CONDUIT
W/ PULL WIRE

TO
JB-1

A-35

TO JB-1

ELEC METER

A-20

DUCT HTR

A-14

JB-10

TO JB-1

W H

#10

A- &G

A-6

#6

A

A-13

#6

COND
UNIT.

JB-1

TO JB-1

JB-1C = 8'-0" AFF TO ℄
OF BOX (TO JB-1C)

A-24

A-27

ELEC. HTRS. BY
OTHERS

WP

SEE SITE PLAN
FOR EXACT LO-
CATION

— POWER

Architect's scales are available in many different degrees or scales; that is, 1 inch = 1 foot, ¼ inch = 1 foot, ⅛ inch = 1 foot, etc. The most common scales are ½ inch = 1 foot, ¾ inch = 1 foot, 1 inch = 1 foot, and ¼ inch = 1 foot. Scales such as ½ inch = 1 foot and ¾ inch = 1 foot are also common on enlarged details.

Figure 5-9 shows part of a building floor plan drawn to a scale of ¼ inch = 1 foot. The dimensions in question are found by placing the ¼-inch architect's scale on the drawing (as shown) and reading the figures. The dimensions read 3 feet, 3 inches.

An architect's scale showing four degrees (⅛ inch = 1 foot, ¼ inch = 1 foot, ½ inch = 1 foot, and 1 inch = 1 foot) is shown in Fig. 5-10. The dimensions of the various lines are as follows: A equals 12 feet, 6 inches on the ⅛-inch scale; B equals 8 feet, 6 inches and C equals 2 feet, 6 inches—both on the ¼-inch scale; D equals 1 foot, 9 inches on the ½-inch scale; E equals 5 feet, 0 inches on the 1-inch scale; and F equals 4 inches on the 1-inch scale.

Every drawing that is prepared in a given reduced scale should be plainly marked as to what the scale is. For example, the building floor plan in Fig. 5-11 should be marked:

FIRST FLOOR PLAN
Scale: ¼ inch = 1 foot 0 inch

ARCHITECTURAL DIMENSIONING

The common rules that apply to other types of drawings also apply to architectural drawings. Only the experienced architectural draftsman can tell, with ease, which dimensions are necessary on an architectural drawing. The dimensions given must be clear and definite, and they must tell the workmen the exact sizes of all parts of the building. Furthermore, they must check with

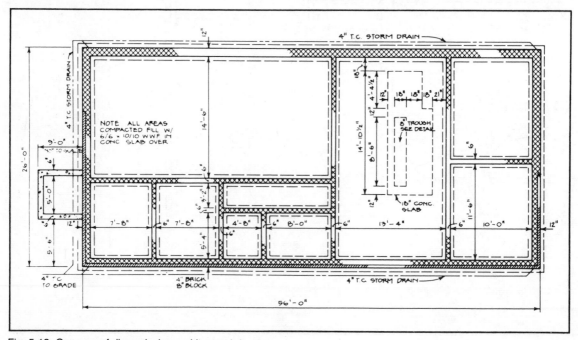

Fig. 5-12. One way of dimensioning architectural drawings.

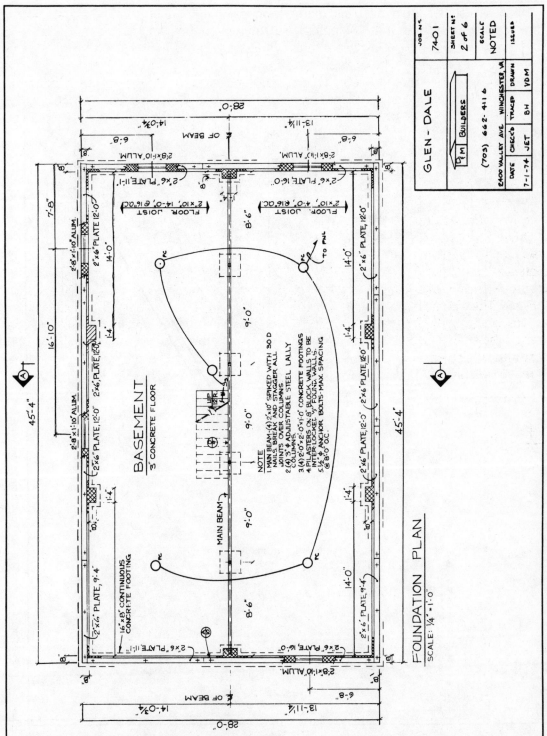

Fig. 5-13. Another way of dimensioning architectural drawings.

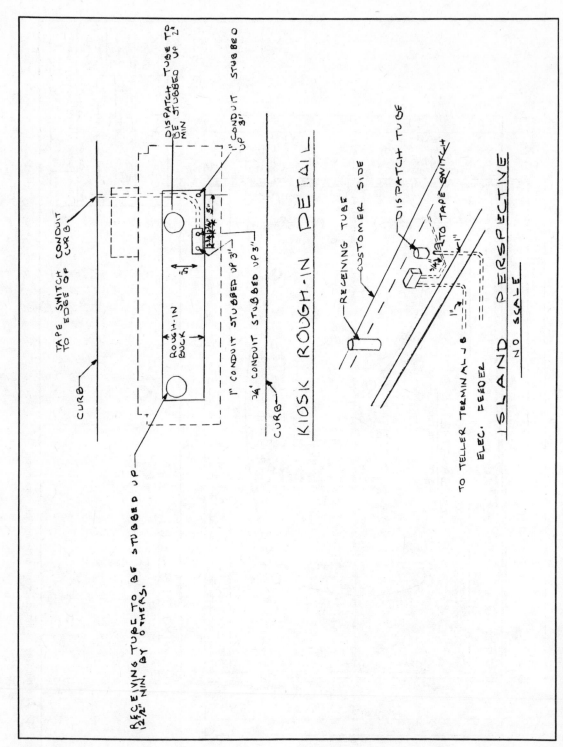

Fig. 5-14. Architectural dimensioning used on electrical drawings.

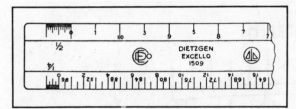
Fig. 5-15. Mechanical engineer's scale.

one another from place to place and from plan to elevation to section, etc.

Several points to be observed in architectural dimensioning are:

•Keep all outside dimension lines well away from the building lines. They should be located a minimum of ¾ inch from the building lines and should be approximately 5/16 to ⅜ inch apart.

•Masonry openings should be marked "M.O." when an exact opening is required. When the opening is for a window or a door, a nominal dimension may be used.

•Dimensions are usually to the centerlines of partitions or else made to the outside walls. The wall thicknesses should also be shown.

•Dimensions should be provided to the centerlines of columns in both directions.

•Dimensions for openings are normally made to the centerline of the opening or else to the sides of the opening as required.

Figures 5-12 and 5-13 show various methods of dimensioning architectural drawings. Figure 5-14 shows the application of architectural dimensioning on an electrical drawing. Notice that most of the dimensions on these drawings are from center to center of various objects. The reason is that lumber and other construction materials often vary slightly in size, making it impossible to indicate actual values of edges of structural members. If structural members vary somewhat in size, the true location is always achieved.

Besides dimension lines and numerals, every drawing should include a plainly marked scale to which the drawing is made. The scale is usually inserted as part of, or adjacent to, the title block or else shown directly beneath a drawing title elsewhere on the sheet.

Anyone who reads working drawings should acquire one or more appropriate scales to determine dimensions of the drawing for items not marked and to check those that are marked on the drawing. The most commonly used scales in the United States are those derived from the common foot rule such as full size, 6 inches = 1 foot, 3 inches = 1 foot, etc. These are the ones most easily read from an ordinary scale, and one of these can usually be adopted. The metric system of measuring is being used more and more, and scales reading directly in the metric system should be considered.

The engineer's scale is the type most often used by land surveyors and for reading site plans on architectural drawings. The use of this scale is fully explained in Chapter 18.

The mechanical engineer's scale is often used in drawing and reading machine and structural drawings. The measuring units on this scale are designed to produce drawings that are to be ⅛, ¼, ½, or full size with the graduations representing inches and fractional parts of inches. Therefore, to read a machine working drawing drawn to ¼ scale, the ¼ measuring face would be used. Each main division on this scale then is equal to 1 inch. The fractional parts of the inch are indicated by small division lines located along the entire edge, or only one unit opposite the zero mark is subdivided. See Fig. 5-15. These scales are available in either flat or triangular shapes.

Chapter 6

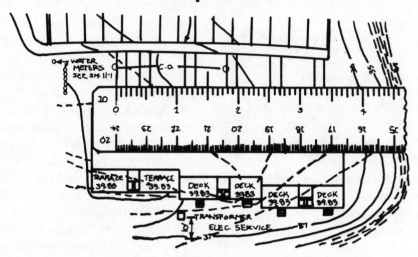

Pictorial Drawings

PICTORIAL DRAWINGS ARE USED EXTENSIVE-ly in many areas requiring instructional material. House plans use more ortho-graphic drawings than any other type. A perspective view of the finished house is sometimes used. Isometric drawings are sometimes used for electrical and plumbing diagrams. Oblique drawings may be used on this same set of plans to indicate the arrangement of, say, the kitchen cabinets.

The hundreds of available electronic kits usually have a pictorial drawing of the circuit board layout along with the conventional schematic drawings. Beginners can thus better understand the placement of the components on the circuit board. Firearm manufacturers frequently supply pictorial drawings (in an exploded view) to aid in the disassembly and assembly of the various weapons. How-to books and articles fre-quently utilize pictorial drawings to show how an object is assembled. Many photos are used, but sometimes it's necessary to see completely around an object—in one view—and a pictorial drawing is the only way to give such a view.

TYPES

Generally, there are three types of pictorial drawings used with instructional material: perspective, isometric, and oblique. The isometric is probably used the most on construction drawings. It can be drawn to scale and the others can't. While all pictorial drawings are useful in helping to present working drawings more clearly, all of them also have disadvantages—namely, intricate parts cannot be pictured clearly and all are difficult to dimension.

Isometric Drawing

Only one view appears in an *isometric drawing* to show several sides. The drawing is formed by using three axes. One axis is vertical and the other two are drawn to the right and left at an angle of 30 degrees to the horizontal (Fig. 6-1). The angle at which the object is drawn (the direction as seen by the viewer) depends entirely on which is the most advantageous side to show.

Isometric drawings may be drawn to scale. The vertical lines (representing height) usually are drawn first. Next come the 30-degree lines representing width and depth of the object being drawn. The necessary surfaces are completed by drawing the necessary lines parallel to the axes. Hidden lines should be omitted in isometric drawing unless they are absolutely essential to describe the object.

Sloping lines that do not run parallel to the isometric axes are called nonisometric lines. Because these lines will not appear in their true length on the drawing, they cannot

be drawn to scale. Remember this when you measure isometric drawings, or else you will have a shape not wanted.

When isometric drawings of irregularly shaped objects are made, the draftsman sometimes uses the *box* construction method. One or more rectangular or square boxes are drawn having sides that coincide with the main faces of the object (Fig. 6-2). The irregular features of the object are then drawn within the framework of the boxes. Again, if you are using isometric drawings to construct an object, remember that only isometric lines are drawn to scale. Don't try to measure nonisometric lines.

Angles usually cannot be shown in isometric drawings to their true shape and size. Such angles are laid off by coordinates drawn parallel to the isometric axes. Isometric circles are normally drawn with isometric templates, but they may be constructed by the four-center system shown in Fig. 6-3. This same four-center method is used to draw isometric arcs, but only the radius

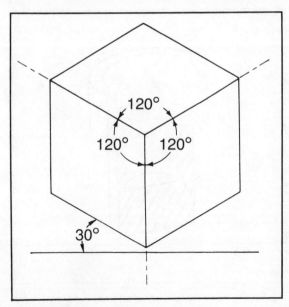

Fig. 6-1. Isometric drawing of a cube.

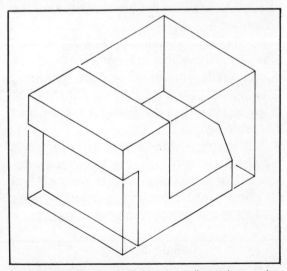

Fig. 6-2. Irregularly shaped objects usually are drawn using one or more rectangular or square boxes with sides that coincide with the main faces of the object when making isometric drawings.

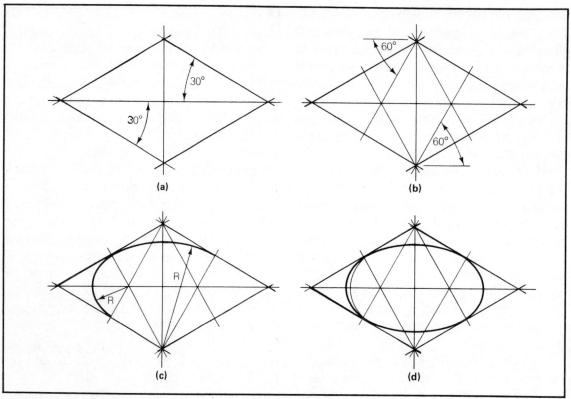

Fig. 6-3. Isometric circles may be constructed by the four-center system.

needed for drawing the arc is used—not the entire circle.

Although an isometric drawing is used mainly to show the exterior of an object, occasionally you may see an isometric drawing illustrating the internal construction of an object (Fig. 6-4). These are known as isometric sections.

Reading dimensions on isometric drawings are similar to the techniques described for multiview drawings. You will find all dimension lines and extension lines parallel to the principal isometric axes.

Oblique Drawing

An *oblique drawing* is one in which the lines of sight are parallel to each other, but the projectors are oblique to the plane of projec-

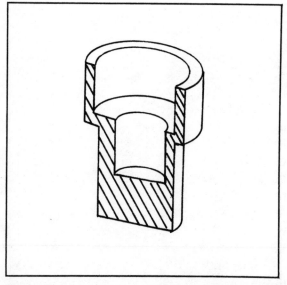

Fig. 6-4. Isometric sections are sometimes used to show the interior of an object.

tion. The principal face is placed parallel to the plane of projection and therefore appears in its true shape. The three kinds of oblique projections used on modern working drawings are: cavalier, cabinet, and general oblique.

Cavalier. This type of drawing is the result of the projectors making any angle of 0 to 90 degrees with the plane of projection. The same scale is used on all axes (Fig. 6-5).

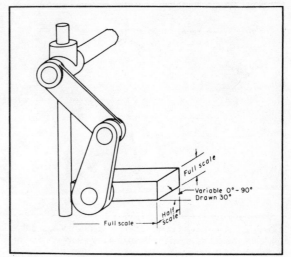

Fig. 6-6. A cabinet projection.

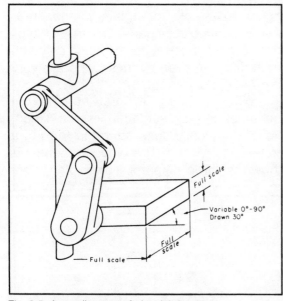

Fig. 6-5. A cavalier type of pictorial drawing.

Cabinet Projection. This type of drawing is one in which the receding lines are foreshortened one-half their actual length to eliminate some of the distortion that is often quite noticeable in cavalier projections. The receding axes are drawn at practically any angle, but they are more commonly found with 30- and 45-degree angles with the horizontal (Fig. 6-6).

General Oblique. A general oblique projection is one in which the projectors make any angle with the plane of projection. The receding axes vary in length from full to one-half scale. The angle of projection is kept between 30 and 60 degrees in most drawings, but the angle used will depend on the shape of the object and the effects desired.

Perspective Drawing

A *perspective drawing* more nearly presents an object as it appears to the eye or is seen in an actual picture. This type of drawing is based on the fact that all lines which extend from the observer appear to come together at some distant point. While viewing an object, the farther away it is, the smaller it seems. Conversely, an object coming closer to the viewer's position seems more and more to approach its true size. An object will appear to the viewer to vary in size, depending upon the distance it is from the viewer.

The *vanishing point* is that point on the horizon where parallel lines, which are perspective lines, seemingly come together and terminate (Fig. 6-7). This point is always on the horizon. The horizon line is always on

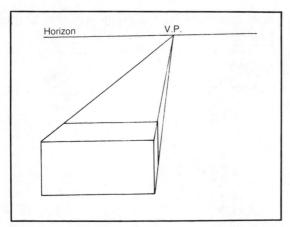

Fig. 6-7. The vanishing point in perspective drawings is that point on the horizon where parallel lines seemingly come together and terminate.

line with the viewer's eyes, and it can change with the viewer's changing position.

Proportion is the comparative relation of one thing to another. In perspective drawing everything such as people, buildings, cars, and the like must be in proportion for the drawing to look correct.

There is very little that can be said about "reading" a perspective drawing, because it is a lifelike representation of the view that should be seen. Notes are sometimes used on technical perspective drawings to convey a message, but these should be self-explanatory.

Perspective drawings can never be drawn to scale. The drawing shown in Fig. 6-8 shows how the object would look if it were actually viewed in "life" form. If you measure the various sides of the object with a scale, you will see that edges AB and CD are not of the same length. If the object itself was measured, they would be the same length. Similarly, the edges EF, EB, and BC, which in this case are also equal on the object itself, will not be the same when measured on this drawing. A perspective drawing is unsuitable for obtaining measurements of an object's parts. The true lengths of lines in perspective can be found only with great difficulty even by those who are very familiar with the method.

When a perspective drawing of an object must be represented in its true dimension, it is usually done so in the form of a projection drawing. The object generally is shown in three different views: plan, front

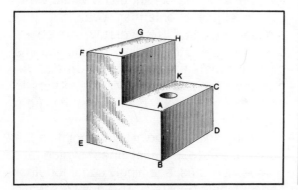

Fig. 6-8. Perspective drawings can never be drawn to scale.

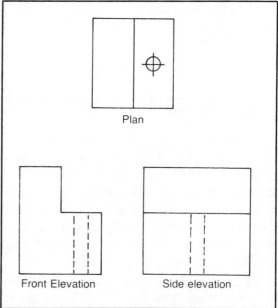

Fig. 6-9. A projection of the perspective drawing shown in Fig. 6-8.

elevation, and side elevation. A projection of the perspective drawing in Fig. 6-8 is shown in Fig. 6-9. Do you understand how these views were obtained?

The object shown in Fig. 6-8 is assumed to be placed within a glass case. The various views take their names from the different positions of the observer in his view of the object through the transparent sides or planes. See Fig. 6-10. Note that the top plane is parallel to the top surfaces of the object. The front plane is perpendicular to the top plane and parallel to the surface ABEFIJ of the object. The side plane is at

right angles to the top plane and also to the front plane.

The three views shown in Fig. 6-10, instead of being shown on three different planes, are represented on one plane as on a sheet of paper. The relation of the views on a single plane will be better understood by assuming that the top plane in Fig. 6-10 is hinged along the front edge, and the side plane is hinged along the vertical edge of the front plane, so that these planes can be swung into the plane of the front view (Fig. 6-11).

The lines on which the planes are

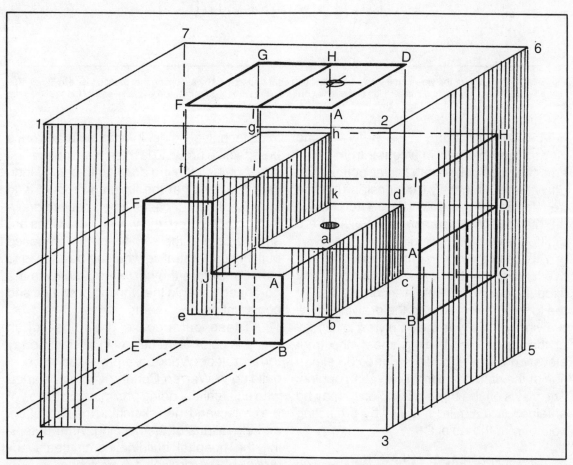

Fig. 6-10. This object is assumed to be placed within a glass case, and the various views take their names from the different positions of the observer through the transparent sides.

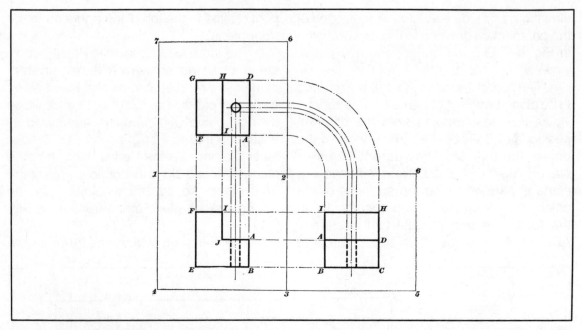

Fig. 6-11. The relation of the views on a single plane will be better understood by assuming that the top plane in Fig. 6-10 is hinged along the front edge, and that the side plane is hinged along the vertical edge of the front plane, so that these planes can be swung into the plane of the front view.

hinged represent the axes between the views. They intersect in a center from which arcs may be scribed to transfer points from one view to another. The dimensions may be transferred by using dividers. The lines of the planes and of the axis of revolution are omitted from the drawing.

To read a drawing of this type, the relative locations of these views must be firmly fixed in your mind. Obtain a correct idea of the form of the objects on the drawing. In all drawings of this type the plan view is usually at the top of the drawing sheet, the front elevation below it, and the right side elevation at the right of the front elevation or plan. The views on the three other planes may be obtained in a similar manner by projecting points from the object.

PRACTICAL APPLICATIONS

Pictorial drawings are used quite extensively in all technical fields. In architectural work a perspective drawing is often used on the first sheet of the working drawing to clearly indicate exactly what the finished product is to look like. In other cases a perspective drawing is made before the working drawings are started so that the owners of the proposed building can visualize what the building is to look like before they give the architects the go-ahead to begin the working drawings and construction documents.

These perspective drawings—often made in color—are helpful in raising money for a project. Work on a proposed hospital building in Warren County, Virginia, started with an architect doing preliminary studies of the project and then sketching the floor plans and elevations of the building. After obtaining the hospital building committee's approval of the preliminary sketches, the architect had a color perspective drawing made of the building that included cars, people, trees, and the like for a realistic appearance. This drawing was then used to show the people of the county what was being done, and they were more willing to contribute to the project.

Pictorial drawings are also used in catalogs to illustrate the objects offered for sale. While many objects are represented in catalogs by photos, some details cannot be readily shown by photography alone. Detailed line drawings are needed. Drawings are also beneficial when a section of an object must be taken to show the object's interior. The object is drawn to appear natural, and then part of the drawing is done

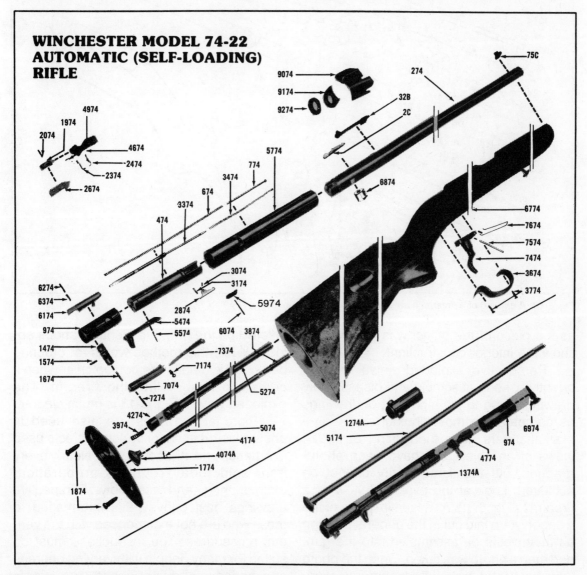

Fig. 6-12. A photographed exploded view of a firearm (courtesy Winchester Repeating Arms Company).

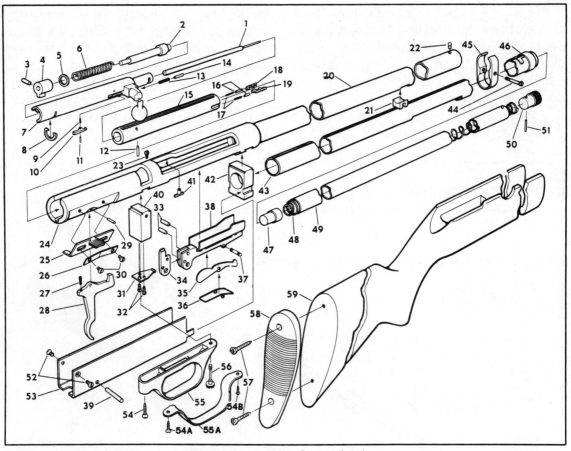

Fig. 6-13. A drawing of a firearm in an exploded view (courtesy Savage Arms).

as if a piece of the object was removed to show the interior construction.

Perspective exploded views are frequently used in assembly and disassembly drawings. Due to the high cost of drawings as compared to most photography, some manufacturers have tried using exploded views of an object that have been photographed, but the same clarity cannot be achieved. Look at the exploded view of a firearm in Fig. 6-12. All essential parts of the weapon were laid out in the general position as they would be assembled into the gun, and then a photo was taken. Once the photo was developed and printed, numerals were lettered on the photo to identify various objects. While this method works for ordering parts, some details are not too clear when it comes to assembling the firearm. The exploded view in Fig. 6-13 is much clearer.

Pictorial drawings are also used in printed instructions for various objects used in America and other countries. Nearly every item sold today requires preparation, maintenance, and assembly. Perspective drawings best convey this information to those who are not experienced with the various procedures. The instructions must be suited for many people with varying degrees of experience and intelligence.

Pictorial drawings used in the industry vary in quality from very excellent and precise to very poor and almost unreadable, unless you have a good understanding beforehand of what is taking place. After gaining some experience, however, you should have little difficulty in reading every pictorial drawing that you will encounter—especially when you have read other drawings described later in this book. If you can read orthographic projections, sections, details, and the like—all requiring a certain amount of visualizing—you should have no trouble reading the average perspective drawing.

Chapter 7

Sectional Views and Details

SOMETIMES THE CONSTRUCTION OF AN OBJECT is so complicated that it is difficult to show in the regular orthographic views. If too many dotted lines are used to show the hidden edges in the object, the views become confusing and hard to read. To show the internal construction clearly, the object is imagined as cut into sections. A simple copper pipe coupling is shown in Fig. 7-1. This simple object really requires no cross section, but it will be used to demonstrate how a section of an object may be taken. Figure 7-2 shows a cross section taken as indicated by the line A-A.

The section in Fig. 7-2 is drawn as though the coupling was cut with a hacksaw down through the center on the line A-A. The arrows indicate the direction from which the section is viewed. Once the cut has been made, the right-hand section is removed so the viewer can see the interior of the left-hand section. If cross-hatching is shown, only the surfaces actually cut by the section line will be crosshatched. In our example the walls of the coupling would be cut with the hacksaw, so only the walls are cross-hatched. The hole itself is left plain.

From the previous example, a section of any object—as applied to working drawings—is what would be seen if the object is sliced or sawed into two parts at the point where the section is taken. Details of the object's construction can be seen. Unless the object to be made is solid and comprises only one material, a sectional drawing must be made to clearly show the details of the object's interior, unless it is constructed of very simple material that can be indicated by dotted lines, notes, or both.

In more complicated projects such as a building or a piece of machinery, working drawings must incorporate sectional views

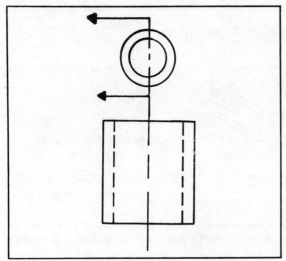

Fig. 7-1. Pictorial drawing of a simple pipe coupling.

taken at various points. Architectural drawings, show sectional views through walls and partitions and through moldings and window trim. Sectional drawings of a complicated machine may show interior cams, holes with the size they are to be threaded, and the like. Anyone who must interpret a working drawing must have a good knowledge of sectional drawings and how to use them.

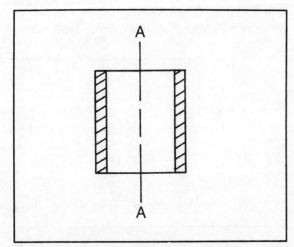

Fig. 7-2. The couple in Fig. 7-1 with one section removed to reveal the interior of the pipe.

EXAMPLES

Considerable imagination must be used when dealing with sectional drawings. Sometimes the problem is very easy, but often it is difficult. There is no set rule for determining exactly what a section will look like; you just have to gain knowledge through experience. Let's consider a few examples.

Sphere and Cube

Any section of a sphere will always be a circle when drawn in an orthographic projection (Fig. 7-3). As long as it's cut clean through, from one side to the other, the resulting section will be a circle when drawn on a flat plane. A cube may have different appearances, depending on where the section is taken. If the cube is sliced from the middle of any side through to the middle of the op-

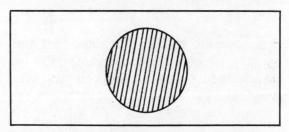

Fig. 7-3. A sphere will always appear as a circle on an orthographic drawing of where the section is taken.

posite side, the resulting section will be a square. If the same cube is sliced slantingly, the resulting section will be a rectangle (Fig. 7-4).

Pear

If a pear is sliced vertically—from top to bottom—down the middle, the section will be pear-shaped. If the same pear is cut horizontally through the middle, the section will be a circle—the diameter of which will de-

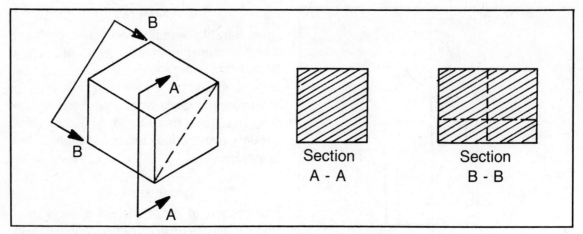

Fig. 7-4. The section of a cube will appear as a square in an orthographic view if cut one way, but it will appear as a rectangle if sliced slantingly.

pend upon where the cut is taken (Fig. 7-5). A piece of round pipe, cut vertically, will appear on the drawing as a rectangle; cut horizontally, it will be a circle; and cut on the slant it will be an ellipse.

Cone

A cone in the position illustrated, cut horizontally, will be a circle; cut slantingly, it will be an ellipse; cut parallel to its slanting side, it will be a parabola; and cut vertically, not through the middle, it will be a hyperbola

(Fig. 7-6). If you understand the basic principles of sections, you will be able to "read" practically any section that can be drawn, especially after gaining some practical experience.

DESIGNATING MATERIALS

In many sectional drawings material types are often indicated by certain combinations of lines. Unfortunately, there is no universally adopted standard at this time. One drawing may have a certain combination of lines to indicate, say, cast iron. This same

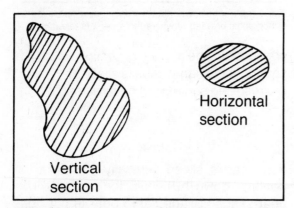

Fig. 7-5. A pear sliced vertically will appear pear-shaped in an orthographic view, but it will appear as a circle if cut horizontally.

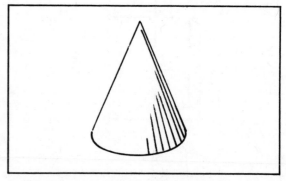

Fig. 7-6. A cone can take several shapes when sectioned at different points.

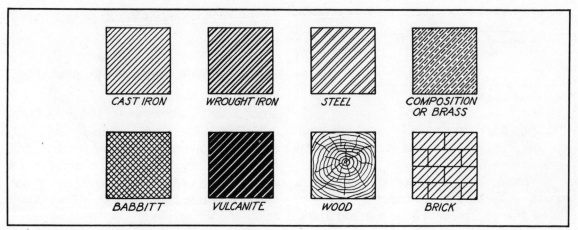

Fig. 7-7. Representations of various materials in sectional views.

combination of lines may indicate brass on another drawing, and so on. There is usually no difficulty experienced due to this diversity of practice because—as a general rule—the material is and should always be distinctly specified on the drawing to prevent any mistake on the part of those people that must read the drawing.

The combinations of lines shown in Fig. 7-7 are the ones most used in this country for designating various materials on working drawings. The name of the material represented by each combination is given below it. The combination of lines that represents cast iron consists of equally spaced lines of uniform thickness. This type of sectioning is frequently used for all materials and will be seen on several types of drawings shown in section.

In architectural drawings it is obvious that the draftsman cannot actually draw every brick and stone in the various walls. To simplify making the drawing, symbols are used (see Chapter 2) to represent the different materials. When you examine the lines or combination of lines used in a building section, note that they are similar to those used to indicate materials used in

steel and other objects, except that they are used to represent building materials.

To indicate brickwork, a series of lines drawn at a 45-degree angle is normally used. The section of the wall is drawn first, and then the space within the outline is cross-hatched or shaded to indicate that the material used to construct the walls is brick.

Concrete walls are indicated in a similar fashion in that the outline of the walls are first drawn and then filled in with little indications of small stones used in concrete. Dots are then placed near the edges to suggest the appearance of concrete.

To indicate stonework on drawings, a method similar to that for concrete is used, except the small stones are omitted. You will find lines drawn at a 45-degree angle used with the dots to indicate stone on some drawings. These lines are similar to those used to indicate brick. The spaces between the lines, however, are usually wider than those used to indicate brickwork.

Terra-cotta is indicated by using the outline and dots similar to those used in indicating stone and concrete, but with lines drawn freehand across the space assumed to be occupied by the terra-cotta. Notice that

Fig. 7-8. Symbols for tile.

the lines are rather irregular, and the small curves and kinks indicate they are drawn freehand.

Hollow tile is indicated by using lines similar to those used to indicate brick, but these are crossed with lines running in the opposite direction.

Plaster is usually indicated by dots, but they are placed very sparingly on the drawing. Notes usually accompany them.

Either of two methods is accepted for showing structural timbers or rough lumber in section. One is to form crossed lines like the letter "X." The other method uses a series of irregular circulars or ovals similar to the grain that shows on the end of a piece of lumber. Sometimes additional lines are used to indicate the checks in timber. The symbols for some building materials are shown in Fig. 1-25.

Finished wood in section is generally indicated by a series of parallel slightly curved lines drawn at an angle of 45 degrees across the space on the drawing.

The symbol for partition tile in Fig. 7-8 is similar to that of hollow tile, except the hatching lines run horizontally and vertically instead of diagonally.

Sheet metal is represented on a drawing by a full heavy solid line. Sometimes attention is called to it via notes.

Cast iron, as discussed previously, is indicated by a series of parallel lines in pairs and at an angle of 45 degrees.

Steel bars and rods, if indicated in section, are represented by solid circles, squares, or rectangles, depending upon their shape.

USING SYMBOLS ON WORKING DRAWINGS

Draftsmen sometimes need to indicate earth around footings or other portions of a building. This is generally done by following around the outline as in Fig. 7-9, which indicates the edge of the portion of earth shown with a series of lines forming a type of border. This border consists of short lines drawn at an angle freehand.

To show how some of the symbols described earlier are used on actual working drawings, look at the section in Fig. 7-10. This is a partial floor plan of a ramp entering an existing building. Note the arrowhead around a circle with the letter "E" and numeral "2." This group of designations indicates that a section is taken through the building as if the view was standing directly on the spot where the arrowhead is. Elsewhere on the actual drawing is the section itself, which appears in Fig. 7-11. Several types of materials are shown in this section. The footings are of concrete as discussed previously. The reinforcing rods are shown using solid circles contained within the footings themselves. Next is a concrete block, upon which is set an 8-inch stone wall. Again, the symbol for concrete is used to show the concrete floor or deck. Wire mesh is used to reinforce this deck as shown by symbol and the note "6 × 6 #10 W.W.F." Other items shown in this section include the steel post, steel balusters and steel handrail molding.

Another section of this same building appears in Fig. 7-12, which is a section of the elevator shaft used in the building. Note the concrete footings, concrete block walls, and a concrete slab covering the shaft. Figure

7-13 is a sill detail showing the concrete slab and connecting angle iron. Note the hatching.

Other details that appeared on this drawing include the yard drain detail shown in Fig. 7-14, section "A" through the building (Fig. 7-15), and millwork details shown in Fig. 7-16. Note the symbols.

DETAILS

The scale at which most working drawings are drawn requires that certain items be expressed in larger size details. Few fields are exempt including electronics, architecture, engineering, machine shop, and the like. To illustrate the use of such details on working drawings, look at a portion of a schematic drawing in Fig. 7-17. From this drawing, an experienced electronic technician should be able to make the project work and devise a method to arrange the components and

conductors. An inexperienced hobbyist may not find the task quite as easy. Heath Company supplemented the schematic diagram with a detail drawing (Fig. 7-18) showing the arrangement of the components on the chassis, how the conductors are tied, etc. The schematic will still have to be used to identify parts and to make the correct connections.

Look at the floor plan of a residence in Fig. 7-19. The drawing shows several lighting outlet boxes installed in a solid wood deck. These will then be ceiling outlets for the area below. It would seem that no further information would be required for the electricians to install the outlets. The deck is constructed of tongue and groove solid lumber, and concealing wiring in this solid deck is a problem. Perhaps the electricians will eventually think of a method to accomplish the task, but a detail drawing such as Fig.

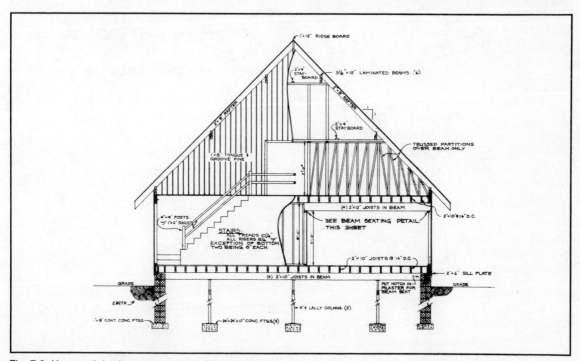

Fig. 7-9. How earth is shown around a building footing.

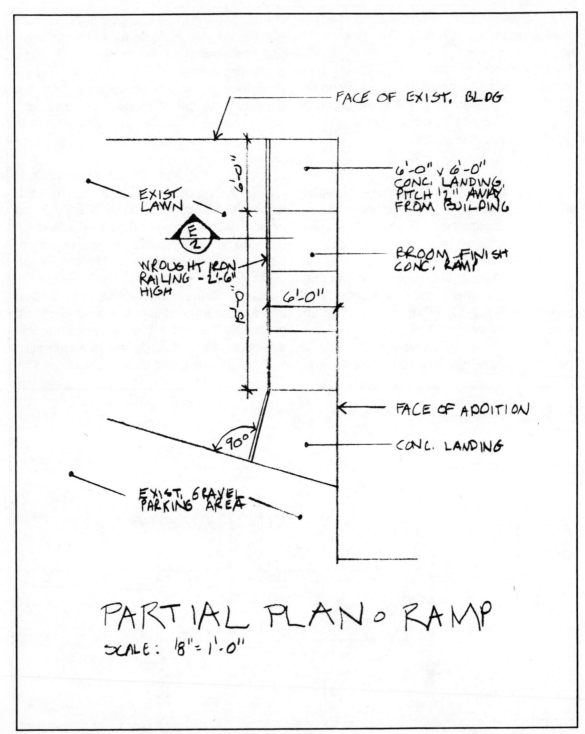

FACE OF EXIST. BLDG

EXIST LAWN

6'-0"

E N

WROUGHT IRON
RAILING - 2'-6"
HIGH

6'-0" x 6'-0"
CONC. LANDING,
PITCH 1/2" AWAY
FROM BUILDING

BROOM FINISH
CONC. RAMP

6'-0"

6'-0"

90°

FACE OF ADDITION

CONC. LANDING

EXIST. GRAVEL
PARKING AREA

PARTIAL PLAN ○ RAMP
SCALE: 1/8" = 1'-0"

Fig. 7-10. Partial floor plan of a ramp entering an existing building.

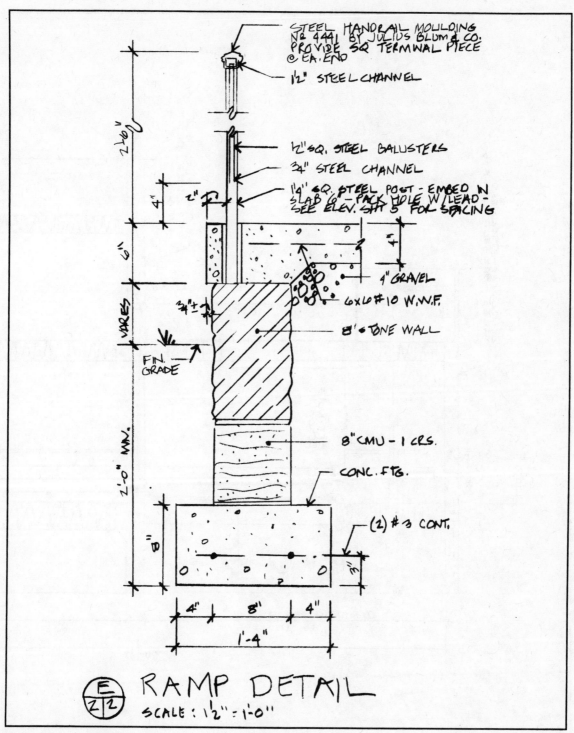

STEEL HANDRAIL MOULDING
No 4441 BY JULIUS BLUM & CO.
PROVIDE SQ TERMINAL PIECE
@ EA. END

1½" STEEL CHANNEL

½" SQ. STEEL BALUSTERS

¾" STEEL CHANNEL

1¼" SQ STEEL POST - EMBED IN
SLAB 6" - PACK HOLE W/ LEAD -
SEE ELEV. SHT 5 FOR SPACING

4" GRAVEL

6 X 6 #10 W.W.F.

8' STONE WALL

8" CMU - 1 CRS.

CONC. FTG.

(2) #3 CONT.

FIN.
GRADE

2'-6"

4"

6"

VARIES

1'-0" MIN.

2"

¾" ±

8"

3"

4"   8"   4"

1'-4"

E
2/2

RAMP DETAIL
SCALE : 1½" = 1'-0"

Fig. 7-11. Ramp detail.

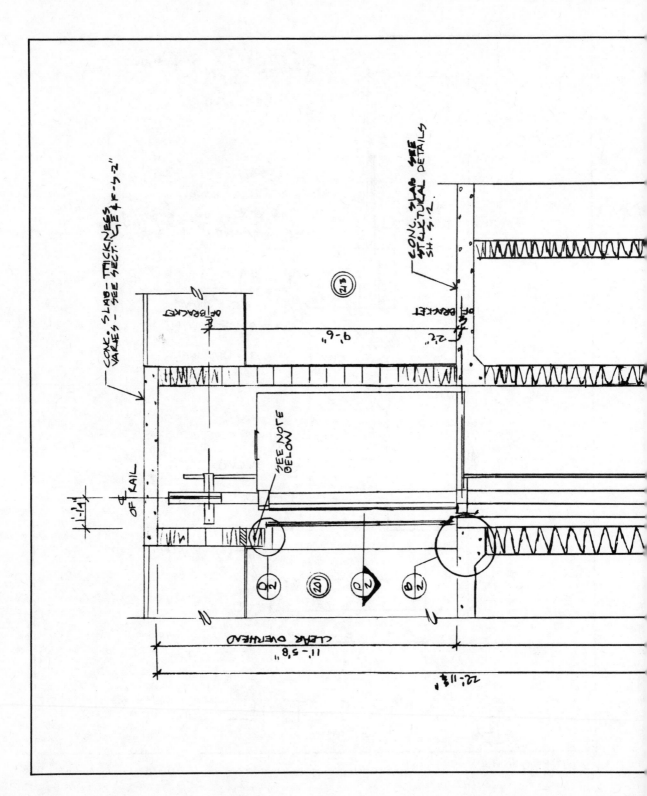

CONC. SLAB THICKNESS VARIES - SEE SECT. 4 SH 5-2"

CONC. SLAB SEE STRUCTURAL DETAILS SH. 5-2.

OF BRACKET

OF BRACKET

9'-6"

2'-2"

SEE NOTE BELOW

OF RAIL

1'-1"

CLEAR OVERHEAD
11'-5,8"

22'-11 3/4"

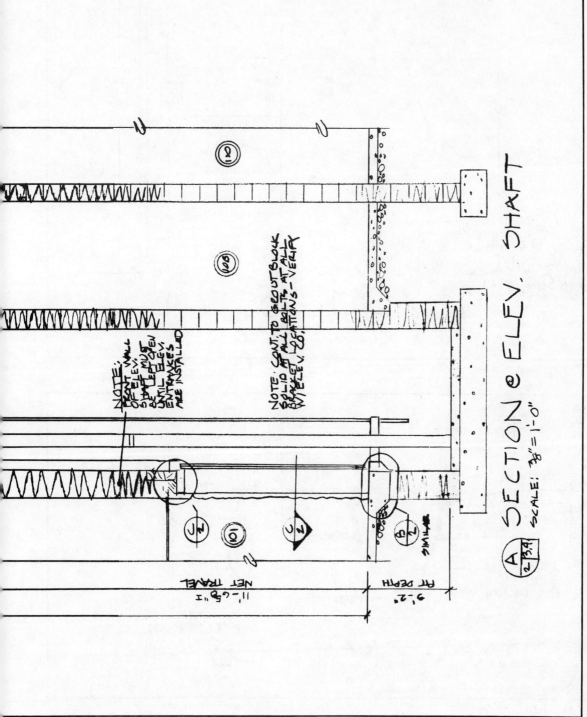

Fig. 7-12. Section of an elevator shaft.

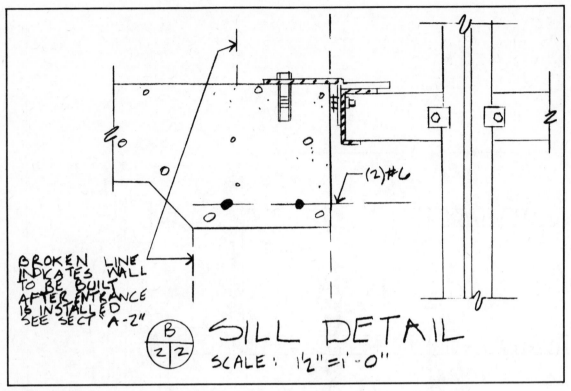

BROKEN LINE INDICATES WALL TO BE BUILT AFTER ENTRANCE IS INSTALLED SEE SECT "A-2"

(2)#6

$\dfrac{B}{2|2}$ SILL DETAIL
SCALE: $1\frac{1}{2}"=1'-0"$

Fig. 7-13. Sill detail.

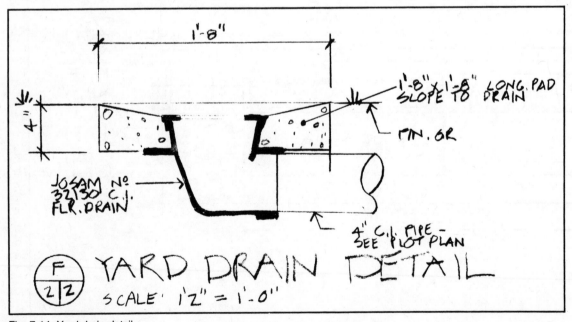

1'-8"

4"

1'-8" X 1'-8" LONG PAD SLOPE TO DRAIN

FIN. GR

JOSAM No 32130 C.I. FLR. DRAIN

4" C.I. PIPE - SEE PLOT PLAN

$\dfrac{F}{2|2}$ YARD DRAIN DETAIL
SCALE: $1\frac{1}{2}" = 1'-0"$

Fig. 7-14. Yard drain detail.

102

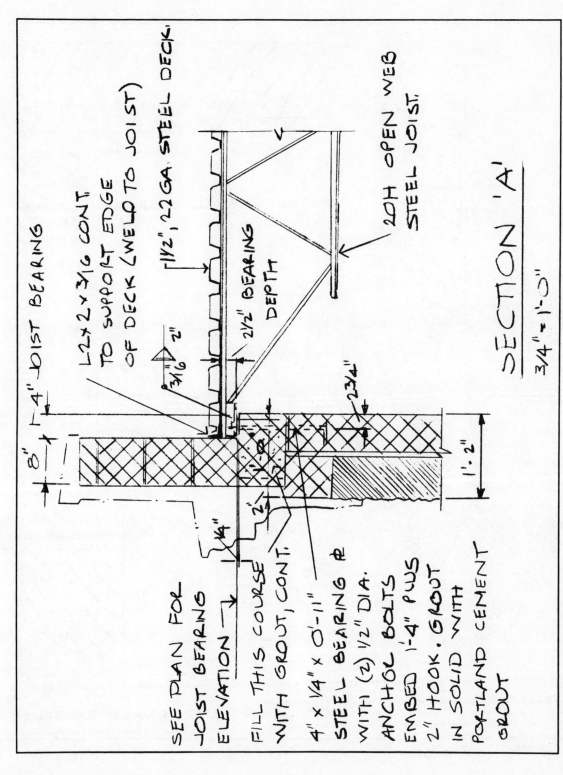

SEE PLAN FOR
JOIST BEARING
ELEVATION

FILL THIS COURSE
WITH GROUT, CONT.

4" x 1/4" x 0'-11"
STEEL BEARING ℞
WITH (2) 1/2" DIA.
ANCHOR BOLTS
EMBED 1'-4" PLUS
2" HOOK. GROUT
IN SOLID WITH
PORTLAND CEMENT
GROUT

1'-4" JOIST BEARING

L2 x 2 x 3/16 CONT.
TO SUPPORT EDGE
OF DECK (WELD TO JOIST)

1 1/2", 22 GA. STEEL DECK.

2 1/2" BEARING
DEPTH

20H OPEN WEB
STEEL JOIST.

8"

3/16 2"

2 3/4"

1'-2"

1/4"

2"

2"

SECTION 'A'
3/4" = 1'-0"

Fig. 7-15. Section "A" through a building.

103

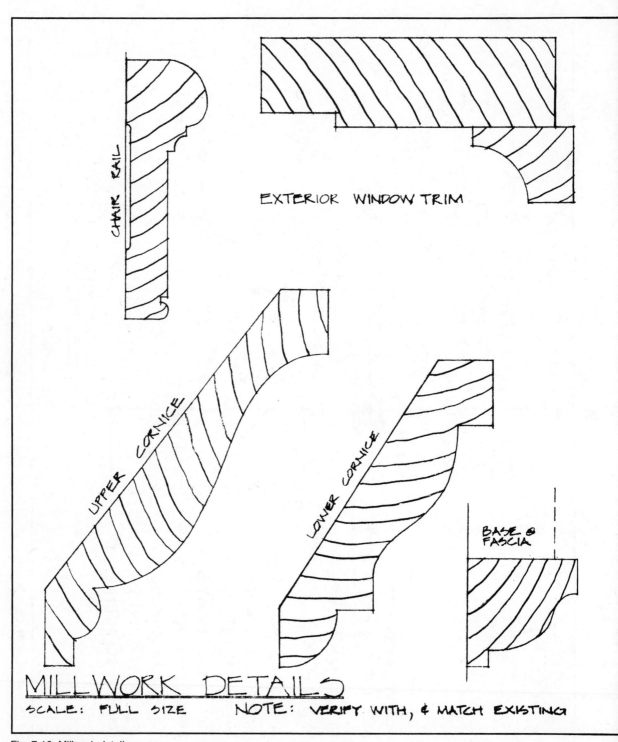

CHAIR RAIL

EXTERIOR WINDOW TRIM

UPPER CORNICE

LOWER CORNICE

BASE & FASCIA

MILLWORK DETAILS
SCALE: FULL SIZE   NOTE: VERIFY WITH, & MATCH EXISTING

Fig. 7-16. Millwork details.

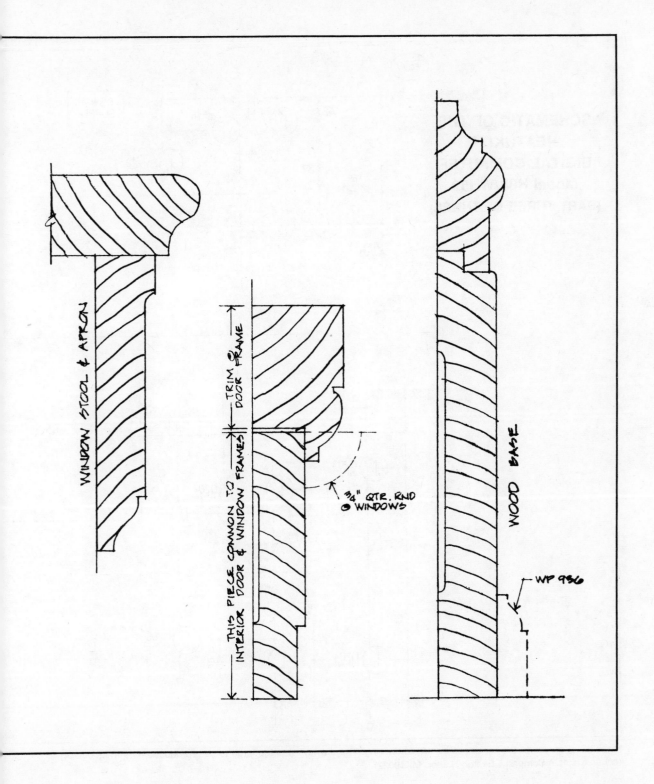

WINDOW STOOL & APRON

TRIM DOOR FRAME

THIS PIECE COMMON TO INTERIOR DOOR & WINDOW FRAMES

3/4" QTR. RND @ WINDOWS

WOOD BASE

WP 936

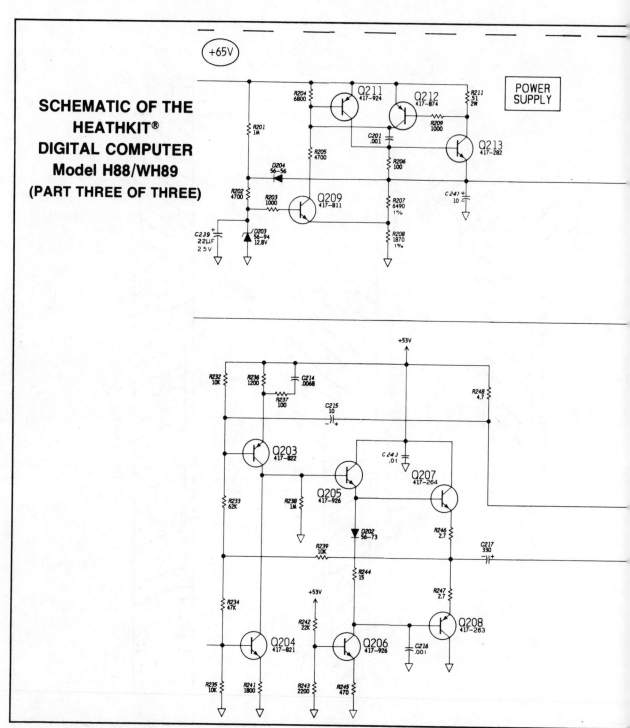

Fig. 7-17. A portion of the schematic electronic drawing. The drawing was too large to be reproduced here in its entirety (courtesy Heath Company, Benton Harbor, MI 49022).

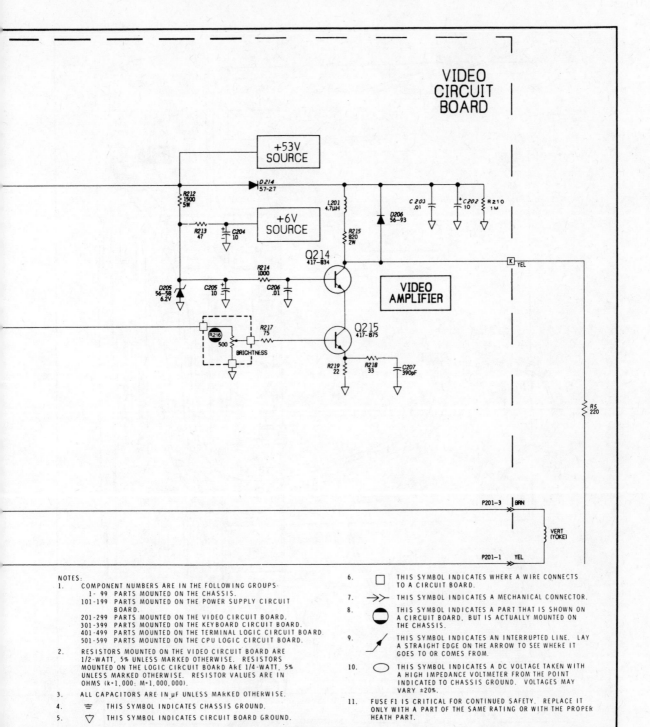

VIDEO CIRCUIT BOARD

+53V SOURCE

+6V SOURCE

VIDEO AMPLIFIER

D214 57-27
R212 1500 5W
R213 47
C204 10
D205 56-58 6.2V
C205 10
R214 1000
C206 .01
R216 500
BRIGHTNESS
R217 75
L201 4.7µH
R215 820 2W
D206 56-93
C203 .01
C202 10
R210 1M
Q214 417-834
Q215 417-875
R219 22
R218 33
C207 390pF
K YEL
R5 220
P201-3 BRN
VERT (YOKE)
P201-1 YEL

NOTES:
1. COMPONENT NUMBERS ARE IN THE FOLLOWING GROUPS:
   1- 99 PARTS MOUNTED ON THE CHASSIS.
   101-199 PARTS MOUNTED ON THE POWER SUPPLY CIRCUIT BOARD.
   201-299 PARTS MOUNTED ON THE VIDEO CIRCUIT BOARD.
   301-399 PARTS MOUNTED ON THE KEYBOARD CIRCUIT BOARD.
   401-499 PARTS MOUNTED ON THE TERMINAL LOGIC CIRCUIT BOARD.
   501-599 PARTS MOUNTED ON THE CPU LOGIC CIRCUIT BOARD.
2. RESISTORS MOUNTED ON THE VIDEO CIRCUIT BOARD ARE 1/2-WATT, 5% UNLESS MARKED OTHERWISE. RESISTORS MOUNTED ON THE LOGIC CIRCUIT BOARD ARE 1/4-WATT, 5% UNLESS MARKED OTHERWISE. RESISTOR VALUES ARE IN OHMS (k=1,000; M=1,000,000).
3. ALL CAPACITORS ARE IN µF UNLESS MARKED OTHERWISE.
4. ⏚ THIS SYMBOL INDICATES CHASSIS GROUND.
5. ▽ THIS SYMBOL INDICATES CIRCUIT BOARD GROUND.

6. ▢ THIS SYMBOL INDICATES WHERE A WIRE CONNECTS TO A CIRCUIT BOARD.
7. ➤➤ THIS SYMBOL INDICATES A MECHANICAL CONNECTOR.
8. ⊖ THIS SYMBOL INDICATES A PART THAT IS SHOWN ON A CIRCUIT BOARD, BUT IS ACTUALLY MOUNTED ON THE CHASSIS.
9. ↗ THIS SYMBOL INDICATES AN INTERRUPTED LINE. LAY A STRAIGHT EDGE ON THE ARROW TO SEE WHERE IT GOES TO OR COMES FROM.
10. ⬭ THIS SYMBOL INDICATES A DC VOLTAGE TAKEN WITH A HIGH IMPEDANCE VOLTMETER FROM THE POINT INDICATED TO CHASSIS GROUND. VOLTAGES MAY VARY ±20%.
11. FUSE F1 IS CRITICAL FOR CONTINUED SAFETY. REPLACE IT ONLY WITH A PART OF THE SAME RATING OR WITH THE PROPER HEATH PART.

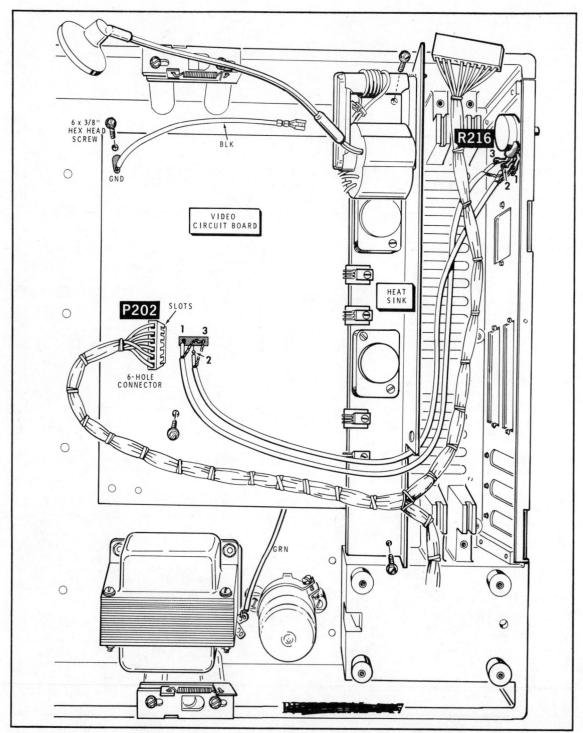

Fig. 7-18. Detail drawing to supplement the schematic in Fig. 7-19 (courtesy Heath Company, Benton Harbor, MI 49022).

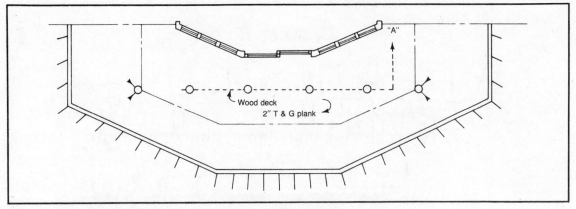

Fig. 7-19. Floor plan of a residence showing location of ceiling outlet boxes.

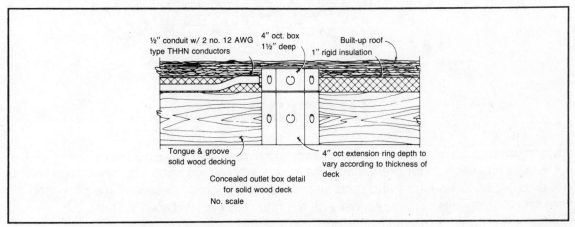

Fig.7-20. Detail drawing used to supplement the floor plan in Fig. 7-19.

7-20 would speed up the work tremendously. With this detail and a few supplemental notes, the installation of the outlet boxes and branch circuits feeding the outlet boxes would involve the following.

Immediately after the wood decking has been installed, all required outlets should be located and marked. A 4-⅛-inch hole saw is then used to drill through the wood deck at each outlet location.

A 4-inch octagonal outlet box with an extension ring is then inserted in the opening; it should be flush with the underside of the deck. Branch circuit conduit feeding each box and strapped against the wood decking will support the box. If only one conduit enters the box (the last outlet on the circuit), a piece of "dummy" conduit about 12 inches long should be connected to the outlet box to provide support for the box on two sides.

THNN conductors should be pulled in the conduit due to the high temperatures expected on the roof from direct sunlight.

When the raceway system is completed, the roofing contractor will install the rigid insulation—leaving channels for the conduit—and then proceed with the built-up roof.

Chapter 8

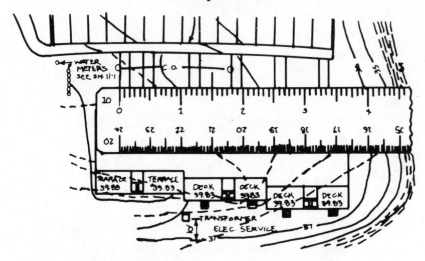

Diagrams

DIAGRAMMATIC DRAWINGS ARE ALSO GREAT time-saving devices for draftsmen. They are also helpful to workmen who must interpret these drawings, as they give—at a glance—the "flow" of the entire system.

Electronic and electrical drawings deal with circuits. They are shown on working drawings by one or more of the following methods:

●Diagrammatic plan views showing individual building-circuit layouts.

●Complete schematic diagrams showing all details of connection and every wire in the circuit.

●One-line diagrams.

●Power-riser diagrams.

Other branches of the industry also use diagrams to a great extent. Computer programs are easier to write once a flow diagram has been made. Plumbing systems depend on flow diagrams to give the desig-ners and workmen an overall view of the system. Air conditioning systems use the same technique to show the air flow and air return of the ductwork system.

COMPUTER FLOWCHARTS

Flowcharts pictorialize the solution to a computer program and are valuable assets to programmers. They are often drawn long before the actual program statements are written.

While flowcharting a program, the approach might change or be simplified, or a flaw might be discovered. After several attempts, a workable flowchart will be the result. The task of programming is greatly reduced once the flowchart is finished.

To learn to read and draw computer flowcharts, you must first learn the basic symbols used. Terminal (starting and ending points) activities are represented by ovals.

Arrows indicate the order of program flow between operations. Most calculator operations are represented by rectangles. A diamond represents a decision point. If the diamond information is computed as "NO," the logical flow continues in another direction.

The following example has each flowchart operation labeled with the corresponding program line number. Once the flowchart is finalized, the program can be written relatively easily (Fig. 8-1).

```
10   REM—THIS PROGRAM AVER-
AGES UP TO 20 POSITIVE NUMBERS.
20   REM—IF YOU HAVE FEWER THAN
20 NUMBERS TO ENTER.
30   REM—ENTER A NEGATIVE NUM-
BER TO END THE INPUTTING.
40   A=0
50   FOR I=I TO 20
60   INPUT B
70   IF B < 0 THEN 110
80   A=A+B
90   NEXT I
110  I=I−1
120  PRINT "THE AVERAGE OF THE "I"
NUMBERS IS "A/I
130  END
```

Flowcharts use boxes and lines, following a few simple rules, to represent a program design. Only a few box shapes and a few ways of drawing the connecting lines are used. These restrictions, allowing only a few flowcharting structures, are intended to make the resulting flowcharts easy to understand by displaying the relationship of parts very clearly. After a correct flowchart has been drawn, translation to a program (writing statements corresponding to the boxes) is relatively simple.

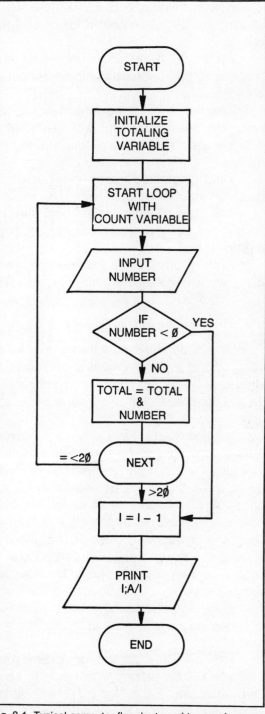

Fig. 8-1. Typical computer flowchart used to organize computer programs.

Flowcharting Symbols. Five different box shapes are usually required for most flowcharts.

Rectangle. A rectangle indicates any processing operation except input/output or a decision.

Diamond. A diamond indicates a decision. The lines leaving the box are labeled with the results that cause each path to be followed.

Parallelogram. A parallelogram indicates an input or output operation.

Oval. An oval indicates the beginning or ending point of a program or program segment.

Circle. A small circle indicates a collection point where lines from other boxes join.

PRACTICAL EXAMPLES

To better understand how these boxes are used, look at a few examples to see how the shapes of boxes are combined with lines to design the processing to be done by a program.

The flowchart in Fig. 8-2 is a very simple program that reads data and prints it. Notice that the flowchart begins and ends with ovals, indicating the start and finish of the program. The sequence of execution (the flow) is always shown by arrows. In this flowchart it is immediately obvious that the actions in the boxes are to be carried out in sequence from top to bottom. Usually the flow of control will not be so simple, but it will always be possible to tell exactly what the intended sequence is by following the arrows.

The purpose of the program represented by this program design is to convert Celsius to Fahrenheit. The parallelogram in the flowchart represents an input of Celsius temperature. The rectangle repre-

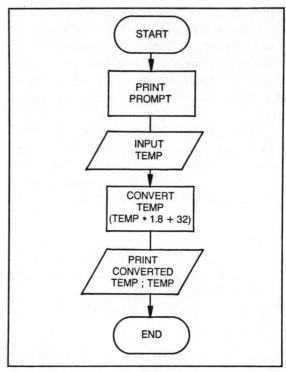

Fig. 8-2. Flowchart used in a temperature conversion program.

sents the processing operation of converting Celsius to Fahrenheit, and the next parallelogram represents the output. The corresponding program follows.

```
10   REM *CELSIUS TO FAHRENHEIT
20   DISP "CELSIUS TEMP"
30   INPUT C
40   LET F=1.8*C+32
50   PRINT C; "C EQUALS"; F; "F"
60   END
```

The program may be easily changed, but using the same flowchart, to convert Fahrenheit to Celsius:

```
10   REM *FAHRENHEIT TO CELSIUS
20   DISP "FAHRENHEIT TEMP";
30   INPUT F
40   C=5/9*(F−32)
50   PRINT F; "F EQUALS"; C; "C"
60   END
```

The fundamental function of a flowchart is to show in exact and unambiguous detail the sequence in which actions are to be carried out, and the conditions under which alternatives are to be taken. In depicting these matters in flowcharts, three control structures are normally used.

Sequence. When an arrow leads from one box to the next, that means simply that the two are to be executed in sequence in the order shown.

Selection. The choice indicated by a decision box always has just two outcomes, depending on whether the condition is true or false.

Iteration. Processing actions are executed repeatedly until a stated condition terminates the repetition. The number of repeats may be zero in some cases.

The basic structures stated earlier may be combined. A path coming out of a decision box may lead to another decision or perhaps an iteration. Furthermore, the actions specified in a loop commonly will include decisions. This is perfectly acceptable so long as only the three basic structures are used.

THE IF...THEN STATEMENT

Often there are times when you want to make a decision. In the averaging program discussed previously we wanted the program to branch to either the end of the program or to the inquiry for more information. The branch was dependent on the outcome of a specified condition, using the IF...THEN statement.

IF numeric expression THEN statement

The IF...THEN statement makes a decision based upon the outcome of the numeric expression. If the expression is true, the THEN part of the statement is exe-

cuted. If the outcome is false, execution continues with the statement following the IF...THEN statement.

Suppose an accountant wants to write a program that will calculate and print the

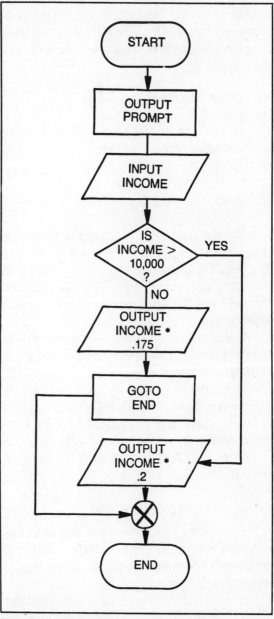

Fig. 8-3. Flowchart used to demonstrate the IF . . . THEN statement.

amount of tax to be paid by several persons. For those with incomes of $10,000 per year or less, the amount of tax is 17.5 percent. For those with incomes of more than $10,000, the tax is 20 percent. A flowchart for the program might look like Fig. 8-3. The diamond in the flowchart would be represented by an IF...THEN statement in a program. A sample program may look like the following:

```
10   DISP "INCOME"
20   INPUT I
30   IF I > 10000 THEN 60
40   PRINT "TAX=";I* .175
50   GOTO 70
60   PRINT "TAX=";I* .2
70   END
```

As you can see, we used a relational operation in the IF...THEN statement. The IF...THEN statement is most often used with relational operators, although the decision can be based on the value of any numeric expression.

If the condition is true—the income is greater than $10,000—then program control is transferred to statement 60. If the condition is false—in this case, if the income is less than or equal to $10,000—then the rest of the IF statement is ignored and the program continues at statement 40. Use the values of $20,000 and $9,000 to test this program.

THE ELSE OPTION

In the previous example, if the numeric expression was evaluated as false, program execution continued with the next sequential statement following the IF...THEN statement. If you specify the ELSE option with the IF...THEN statement, the program will instead perform the indicated ELSE instruc-

tions. This gives you tremendous power with conditional branching; six different forms of the IF...THEN statement are available. The general format of the IF...THEN...ELSE statement is:

IF numeric expression THEN state number or executable instructions. ELSE statement number or executable instructions.

If the numeric expression is false and ELSE is specified, execution is transferred to the line number following the ELSE, or the indicated ELSE statement is executed. Look at the following example.

A quadratic equation is of the form $0=ax^2+bx+c$. If $a \neq 0$, its two roots may be found by the formulas:

$$r_1 = \frac{-b+\sqrt{b^2 - 4ac}}{2a}$$

$$\text{and } r_2 = \frac{-b-\sqrt{b^2 - 4ac}}{2a}$$

To write a program to compute the roots of a quadratic equation given the values of the coefficients a, b, and c, proceed as follows. If a is zero, display an error message and re-enter new values. If $b^2 - 4ac$ is less than zero, then the square root of that value would give a warning message or an error. Make sure that b^2-4ac is greater than or equal to zero before you compute the roots. The flowchart is shown in Fig. 8-4. In this solution we use two forms of the IF . . . THEN . . . ELSE statement. Study it carefully and then load the program and run it.

```
10   REM *ROOTS*
20   DISP "IF A QUADRATIC"
30   DISP "EQUATION IS OF THE"
40   DISP "FORM 0=A*X 2+B*X+C"
50   DISP "ENTER A,B,C"
60   INPUT A,B,C
70   D=B*B−4*A*C
```

```
80   IF A=0 THEN DISP "A=0; NOT
     QUADRATIC.
     REENTER VALUES" ELSE 100
90   GOTO 60
100  IF D>=0 THEN 120 ELSE DISP
     "COMPLEX ROOTS:
     CANNOT COMPUTE. REENTER
     VALUES."
110  GOTO 60
120  R1=(−B+SQR(D))/2∗A)
130  R2=(−B−SQR(D))/2∗A)
140  PRINT "COEFFICIENTS=";A;B;C
150  PRINT "ROOTS=";R1;R2
160  END
```

To summarize, if A=0, the program displays message, then continues to the next statement. If A=0, the program reads ELSE 100 and then branches to statement 100. If D>=0, the program branches to statement 120. If D<0, the program displays an ELSE message and then continues to statement 110.

Run the program to find the roots of the equation $x^2+x−6=0$. Then run the program again to test the decision with $x+1=0$ and $x^2+2x−1=0$.

```
RUN
IF A QUADRATIC
EQUATION IS OF THE
FORM 0=A*X 2+B*X+C
ENTER A,B,C
?
1,1−6 (x²+x−6)
COEFFICIENTS =1 1−6
ROOTS=2−3
RUN
IF A QUADRATIC
EQUATION IS OF THE
FORM 0=A*X 2+B*X+C
ENTER A,B,C
?
-1,2,2(X²+2X+2)
```

```
COMPLEX ROOTS: CANNOT COMPUTE
REENTER VALUES
?
3,2,−1 (3X²+2X−1)
COEFFICIENTS=3 2−
ROOTS=.333333333333−1
```

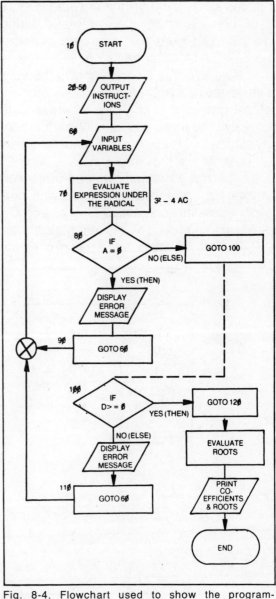

Fig. 8-4. Flowchart used to show the programming.

115

ELECTRICAL DIAGRAMS

Complete schematic *electrical wiring diagrams* are used in highly unique and complicated electrical systems such as control circuits. Components are represented by symbols. Every wire is either shown by itself or included in an assembly of several wires that appear as one line on the drawing. Figure 8-5 shows a complete schematic wiring diagram for a three-phase, ac magnetic nonreversing motor starter.

Note that this diagram shows the various devices in symbol form and indicates the actual connections of all wires between the devices. The three-wire supply lines are indicated by L_1, L_2, and L_3. The motor terminals of motor M are indicated by T_1, T_2, and T_3. Each line has a thermal overload-protection device (OL) connected in series with normally open line contactors C_1, C_2, and C_3, which are controlled by the magnetic starter coil, C. Each contactor has a pair of contacts

that close or open during operation. The control station, consisting of start push button 1 and stop push button 2, is connected across lines L_1 and L_2. An auxiliary contactor C_4 is connected in series with the stop push button and in parallel with the start push button. The control circuit also has normally closed overload contactors (OC) connected in series with the magnetic starter coil (C).

Single-Line Diagrams

Figure 8-6 shows a typical *single-line diagram* of an industrial power distribution system. In analyzing this diagram, the utility company will bring its lines (from 22.9 to 138 kV) to a substation outside the plant building. Air switches, lightning arresters, single-throw switches, and an oil circuit breaker are provided there. This substation also reduces the primary voltage to 4160 volts by transformers. Again, lightning arresters and various disconnecting means are shown.

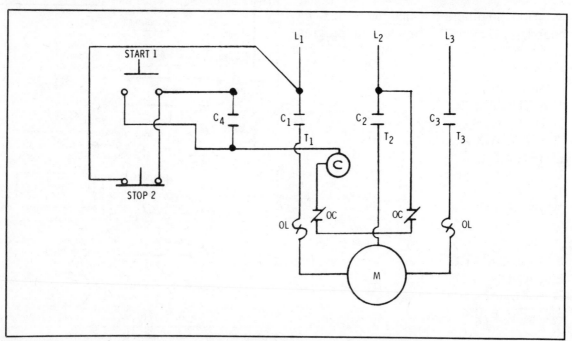

Fig. 8-5. Schematic diagram of a three-phase motor control, including push-button stations.

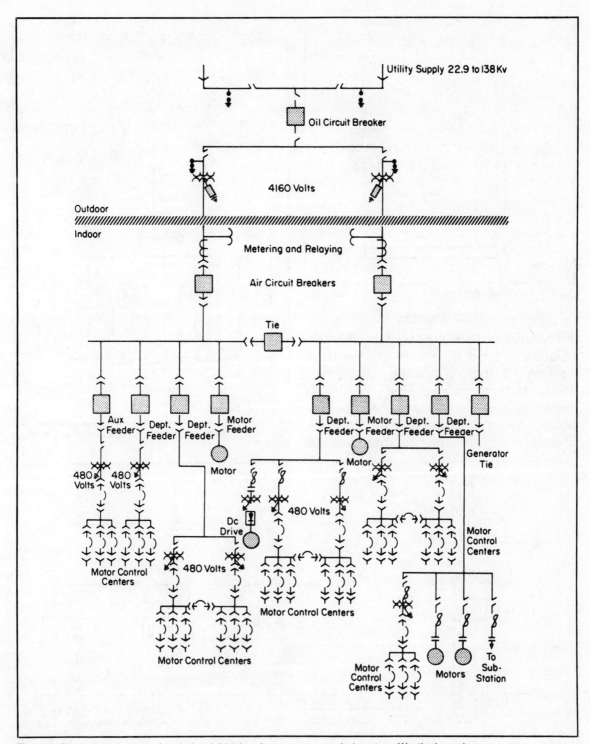

Fig. 8-6. Single-line diagram of an industrial high voltage power supply (courtesy Westinghouse).

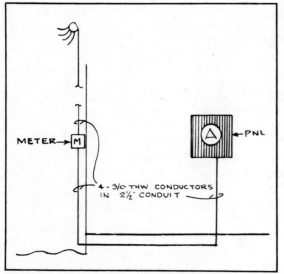

Fig. 8-7. Power-riser diagram of a typical electric service.

Riser Diagrams

Power-riser diagrams are probably the most frequently used diagrams on electrical working drawings for building construction. These diagrams indicate what components are to be used and how they are to be connected in relation to one another. This type of diagram is easily understood and requires

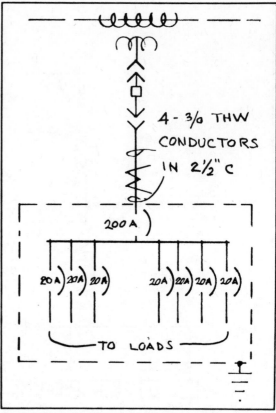

Fig. 8-8. Single-line diagram of the same service as shown in Fig. 8-7.

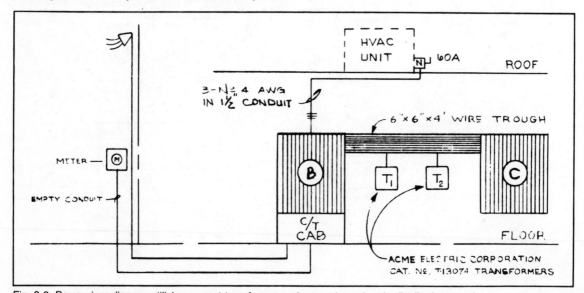

Fig. 8-9. Power-riser diagram utilizing current transformers, wire trough, and main distribution panels.

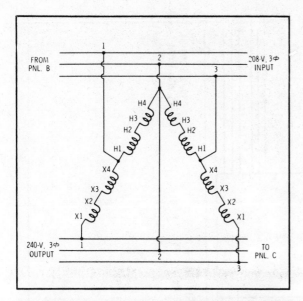

FROM
PNL. B

208-V. 3φ
INPUT

H4 H4
H3 H3
H2 H2
H1 H1

X4 X4
X3 X3
X2 X2
X1 X1

240-V. 3φ
OUTPUT

TO
PNL. C

Fig. 8-10. Schematic diagram of a typical transformer connection.

much less time to draw or interpret than the previously described diagrams. As an example, compare the power-riser diagram in Fig. 8-7 with the schematic diagram in Fig. 8-8. Both are diagrams of an identical electrical system, but the drawing in Fig. 8-7 is greatly simplified.

The power-riser diagram in Fig. 8-9 was used on a working drawing for a printing company building. The main building service was to be three-phase, four-wire, 120/208-volt connected. Several old single-phase motors rated at 240 volts were to be used in some printing machines. Because running these motors on two phases of the 208-volt service would greatly shorten the life of each 240-volt motor, it was decided to specify an additional panel (C) for use with the 240-volt equipment. This panel would connect to two booster transformers "Y-Δ," as shown in Fig. 8-10, to gain the additional voltage.

The power-riser diagram in Fig. 8-11 combines a single-line diagram with a con-

ventional "block" riser diagram to convey the information. It is clear that the pad-mounted transformer, current transformers (CT), and watt-hour meter are furnished by the utility company (VEPCO). The feeder conductors from the transformer to the main distribution panel are also furnished by the utility company. The electrical contractor is required to furnish six 4-inch conduits for the utility company's conductors.

ELECTRONIC DIAGRAMS

Electronic diagrams are used in a similar manner as electrical power diagrams. Some are schematics, showing the various components by symbol, with inner connecting lines as shown in Fig. 8-12. Other examples of electronic diagrams include single-line diagrams such as Fig. 8-13. Pictorial diagrams are used quite frequently, especially in electronic kits for do-it-yourselfers. See Fig. 8-14.

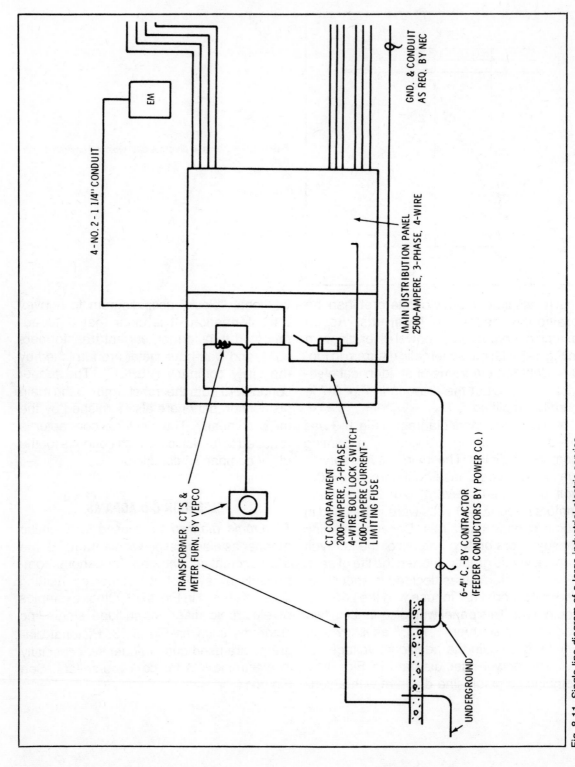

Fig. 8-11. Single-line diagram of a large industrial electric service.

EM

4 - NO. 2 - 1 1/4" CONDUIT

TRANSFORMER, CT'S &
METER FURN. BY VEPCO

MAIN DISTRIBUTION PANEL
2500-AMPERE, 3-PHASE, 4-WIRE

GND. & CONDUIT
AS REQ. BY NEC

CT COMPARTMENT
2000-AMPERE, 3-PHASE.
4-WIRE BOLT LOCK SWITCH
1600-AMPERE CURRENT-
LIMITING FUSE

6-4" C. -BY CONTRACTOR
(FEEDER CONDUCTORS BY POWER CO.)

UNDERGROUND

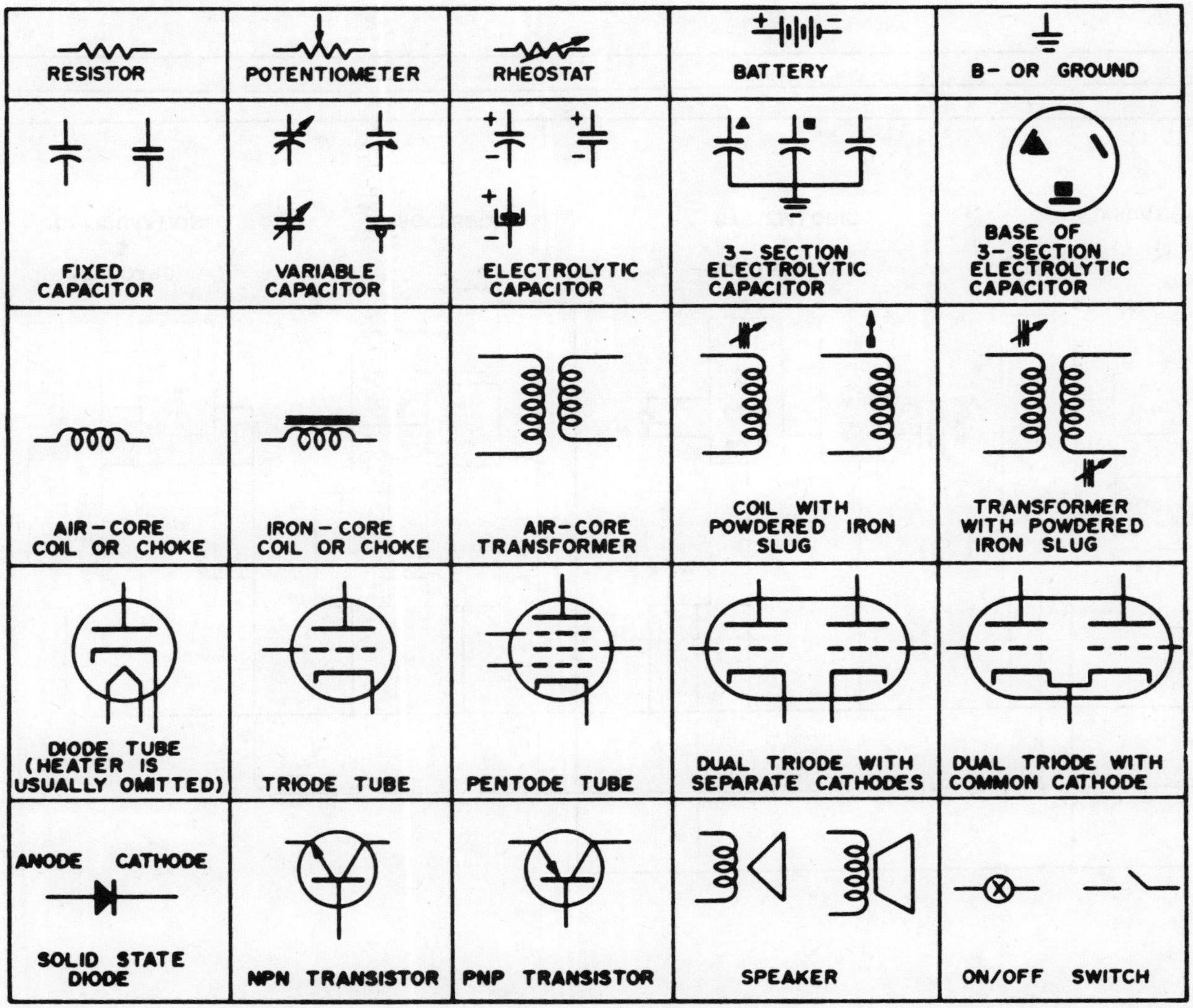

Fig. 8-12. Electronic schematic drawings are made with lines and symbols to represent the various components.

The symbols shown include:

- RESISTOR
- POTENTIOMETER
- RHEOSTAT
- B— OR GROUND
- BATTERY
- FIXED CAPACITOR
- VARIABLE CAPACITOR
- ELECTROLYTIC CAPACITOR
- 3—SECTION ELECTROLYTIC CAPACITOR
- BASE OF 3—SECTION ELECTROLYTIC CAPACITOR
- AIR—CORE COIL OR CHOKE
- IRON—CORE COIL OR CHOKE
- AIR—CORE TRANSFORMER
- COIL WITH POWDERED IRON SLUG
- TRANSFORMER WITH POWDERED IRON SLUG
- DIODE TUBE (HEATER IS USUALLY OMITTED)
- TRIODE TUBE
- PENTODE TUBE
- DUAL TRIODE WITH SEPARATE CATHODES
- DUAL TRIODE WITH COMMON CATHODE
- SOLID STATE DIODE (ANODE / CATHODE)
- NPN TRANSISTOR
- PNP TRANSISTOR
- SPEAKER
- ON/OFF SWITCH

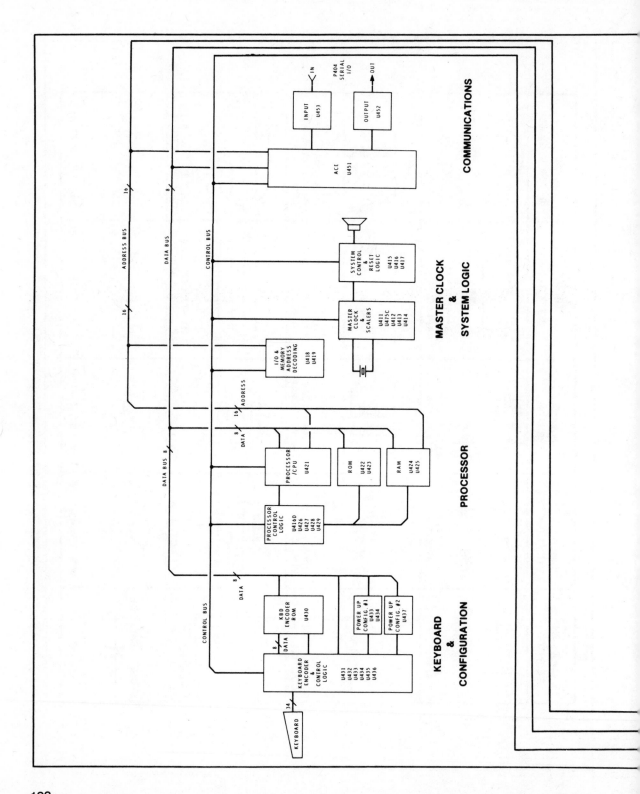

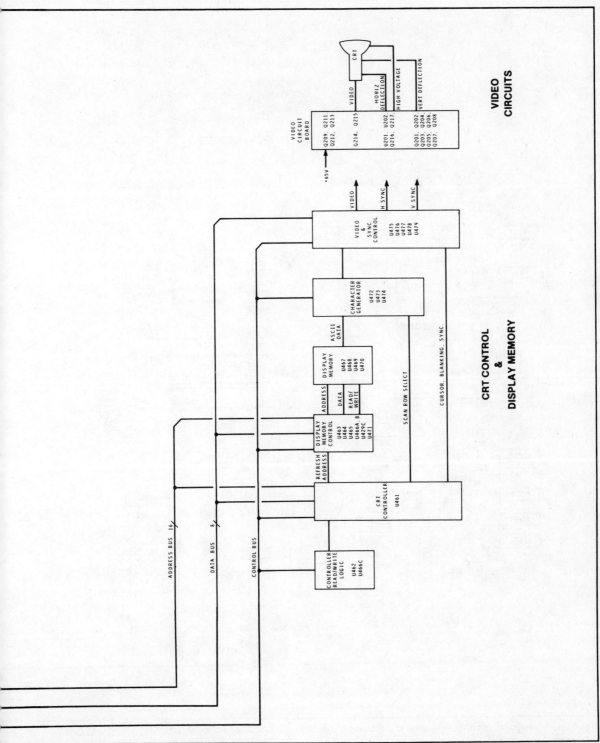

Fig. 8-13. Typical single-line electronic diagram (courtesy Heath Company, Benton Harbor, Michigan).

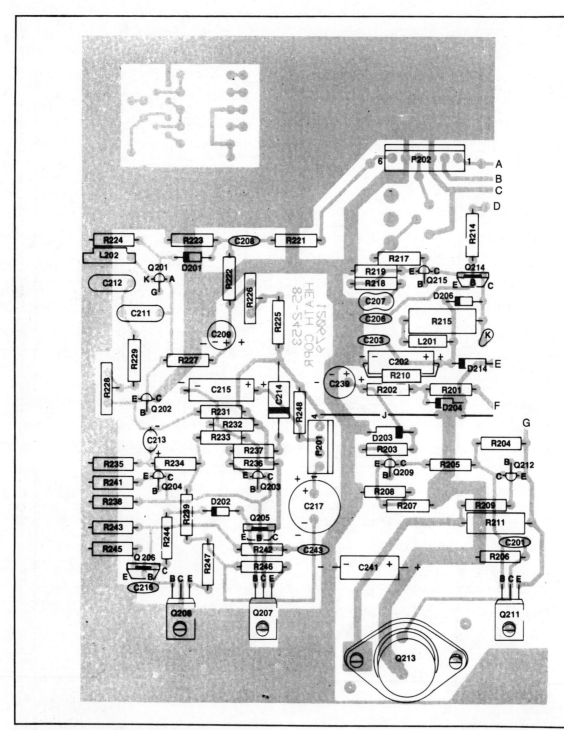

Fig. 8-14. Pictorial diagrams are frequently used in electronic kits designed to be built by hobbyists (courtesy Heath Company, Benton Harbor, MI 49022).

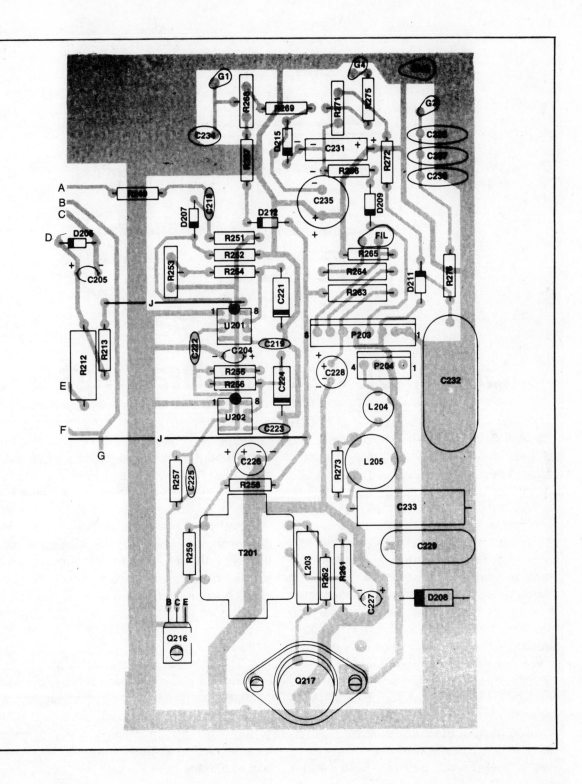

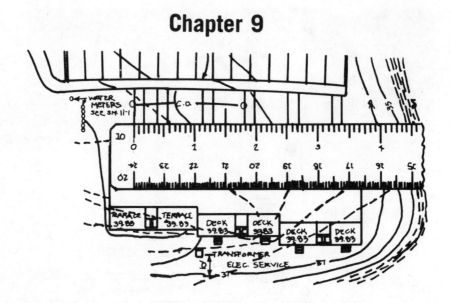

Chapter 9

Layout of Working Drawings

WORKING OR SHOP DRAWINGS ARE USED BY workmen in making tools, objects, and the like. They are drawn to scale and must show all necessary details so the workmen can successfully accomplish their jobs.

Because solid objects all have three dimensions, six views are usually required to show the object completely—the top, the bottom, and the four sides. This number of views is not always necessary. A working drawing should be accurately drawn to scale showing as many views as necessary so that the object can be accurately constructed. Once the required amount of views (and other information) has been given, no further information should be included. It will only add to the confusion and be of no use to the workmen.

The number of views necessary to construct an object or a system depends entirely on the project's complexity. There can be any number of drawings. If you wanted to construct a croquet ball from yellow pine, there would be no need to show more than one view. All views would be alike no matter from which direction the ball was viewed. If drawn to scale and with a note stating "constructed from solid yellow pine," any qualified workman should be able to construct the ball without further information. If even a simple shed is to be constructed, a plan view, elevations of each side of the shed, perhaps a structural drawing, and some sections through the building will be required.

The number and types of drawings required on a project depend entirely on how complex the project is and, to some extent, on who will be using the drawings. The schematic diagram of an electronic metal locator may be all that is required for a person experienced in building electronic proj-

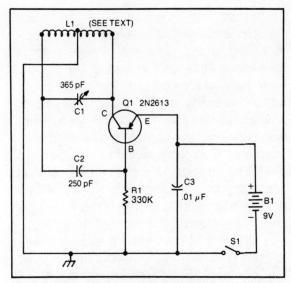

Fig. 9-1. Schematic diagram of an electronic metal locator.

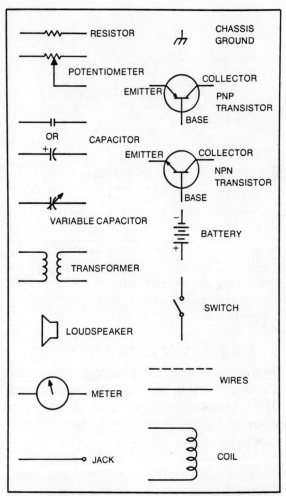

Fig. 9-2. Typical symbol list that accompanies most electronic working drawings.

ects (Fig. 9-1). The beginner will need additional information. A symbol list, like the one shown in Fig. 9-2, should be provided so that the person "reading" the drawings will know for what each symbol stands. A parts list also will be helpful. A pictorial drawing, like the one in Fig. 9-3, will help the builder lay out the parts or components on the circuit board. Finally, notes should be provided to caution the builder against short circuits and how to turn the metal locator once it is built.

A working drawing for a machine shop project may appear as Fig. 9-4—a plan and elevation view. This crank could be constructed in two ways. One way is to make a mold and then cast the part, or else take a solid piece of cast iron and machine the dimensions as called for on the drawing. Although this looks like a simple project, the workmen must be experienced to construct the crank as called for on the drawing.

Another drawing (this time in an isometric view) is shown in Fig. 9-5. This drawing alone is sufficient for experienced workmen, but the apprentice or hobbyist would more

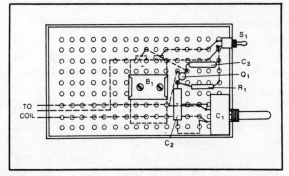

Fig. 9-3. Pictorial layout of electronic components on a circuit board.

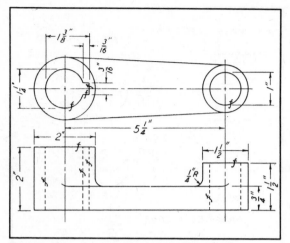

Fig. 9-4. Plan and elevation view of a crankshaft.

than likely need supplemental information concerning construction of an automotive spring bracket.

BUILDING CONSTRUCTION DRAWINGS

Let's analyze a set of building construction drawings to get a "feel" of how drawings are laid out. Although other types of drawings will differ slightly, the same general procedure is used.

The first page of a set of building construction drawings is usually the title page containing the name and location of the building project, the name of the owners and the architect, and any consulting engineering firms used. An index of the drawings sheets is normally provided, along with a legend or key for materials used in the drawing sheets to follow. Drawings will vary, but this is the way most are drawn at this time. See Fig. 9-6.

The second page of the construction drawings will sometimes have a perspective view of the building in which the set of drawings cover. It may have a drawing of the building site. A breakdown of soil borings taken to examine the bearing qualities of the

earth may also be indicated on this sheet.

On smaller projects utility services—including water, gas, telephone, electric, television, etc.—are indicated on the plot plan that usually occupies the space on the second drawing sheet. On very large projects a separate sheet titled "Utility Plot Plan" will be included to facilitate the reading of the drawings. Each utility will have its own plot plan. The underground or overhead electric lines may be so complex that they will appear on a separate drawing of a plot plan and show only electrical wires and services.

Drawing sheets immediately following the plot plan will normally contain the building floor plans. Examples appear in Fig. 9-7. Some sheets may also contain cross sections of the building with the sections drawn on the same sheets (in the margins) as the

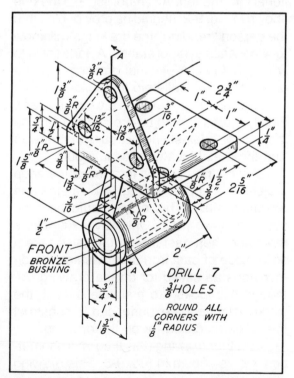

Fig. 9-5. Isometric working drawing that gives all necessary construction details.

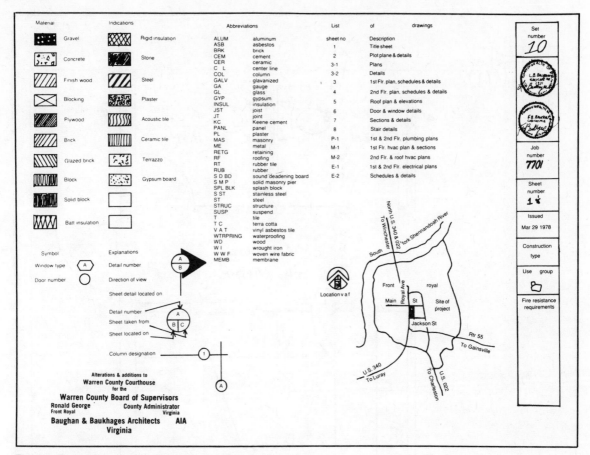

Fig. 9-6. Typical title sheet found on most architectural drawings.

floor plans. Otherwise, sections are indicated on the floor plan sheets, but the sections themselves are drawn on another sheet or sheets. See Fig. 9-8.

Elevations are also provided where space is available—either on the same drawing sheets with the floor plans or on separate sheets. Again, the size of the project will dictate the number of drawings and sheets required.

These architectural drawings will be followed by structural, mechanical, and electrical drawings as required. Written specifications almost always accompany a set of architectural drawings and outline the grades of material, workmanship, and similar data.

GENERAL INFORMATION

Before any drawing is begun, a preliminary study is usually made to determine what views will be necessary to convey the required information and how they will be arranged. The size of the drawings are also determined at this time. An 8½ × 11-inch sheet of drawing paper may be sufficient, or the drawings may require a sheet 24 × 36 inches or larger.

In preparing working drawings a heavy border is normally placed about ½ inch in-

129

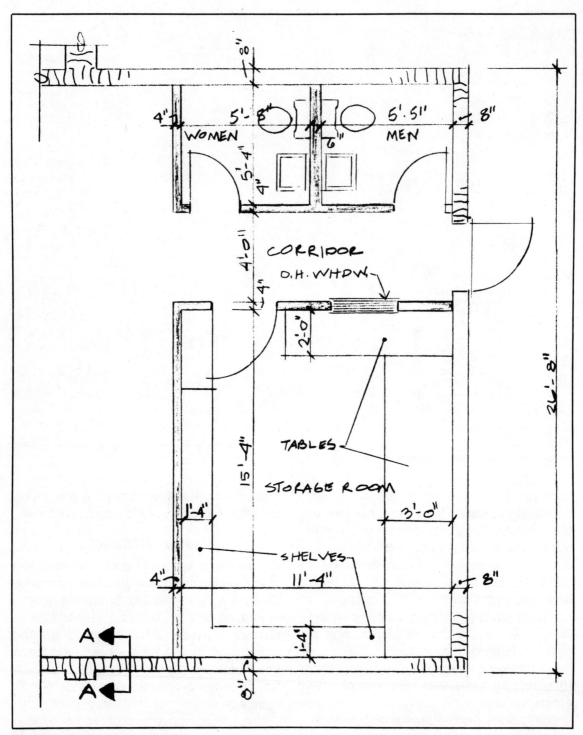

Fig. 9-7. Partial floor plan of a building.

side the outside perimeter of the drawing sheet. Space for a title block is also provided to identify the drawing.

A decision is made as to what views are necessary for the drawing and how many drawing sheets will be required. The scale to which the views are drawn is very important and is usually decided on before the drawing commences. Where dimensions must be held to extreme accuracy, the scale drawings should be as large as practical with dimension lines added. Where dimensions require only reasonable accuracy, the object may be drawn to a smaller scale (with dimension lines possibly omitted). The object can be scaled with an architect's, engineer's or mechanical engineer's scale.

When a project is being drawn, a designer almost always plans and designs the object. The draftsman then transforms the designs into working drawings. The following usually takes place:

●The engineer or designer determines how the design should be conveyed.

●A rough sketch or sketches are made to obtain the best possible layout.

●Any required calculations are made.

●Schedules are roughly laid out for the draftsman.

●Any required diagrams are also made.

●A legend or symbol list is indicated.

●Various large-scale details and section views are roughly sketched. If necessary, to show exactly what the workmen must do.

●Written specifications are then made, if necessary, to give a description of the materials and installation methods.

●The draftsmen, under the designer's supervision, then prepares the working drawings.

●When the drawings are complete, the designer (and possibly other checkers) reviews the drawings for accuracy.

SADDLE ADJUSTING LEVER

An assembly drawing of a saddle adjusting lever is shown in Fig. 9-9. This lever is used on a 5-foot boring mill and is indicated on the drawing in the upper right-hand corner. Note that a border is drawn all around the title blocks and the drawing itself. To assemble this device, only plan and elevation views were needed. Various parts necessary for the assembly of the device are numbered on the drawing with arrowheads pointing to the respective part. The numbers again appear on a component list so that each part can be identified.

Other information on this drawing includes the group number, the date the drawing was made, and the amount or number of levers required.

SYMBOLS

We have previously discussed using symbols and other time-saving devices when making working drawings. Usually these symbols are used in diagrammatic form to be "read" by experienced workmen. Even full-sized detailed drawings sometime utilize simplified techniques to shorten the time required to make the drawing. Two forms of symbols for representing threads on a screw or bolt are shown in Fig. 9-10. The regular form shown in Fig. 9-10A consists of a series of alternate long and short lines drawn at right angles to the axis. These lines represent the crests and the roofs of the threads. In most cases the short lines—representing the roots—can be made heavier than the long lines. To save drafting time, the short form (simplified form) shown in Fig. 9-10B is used. The threaded portion is indicated by lines made of short dashes drawn parallel to the axis at the approximate depth of the thread.

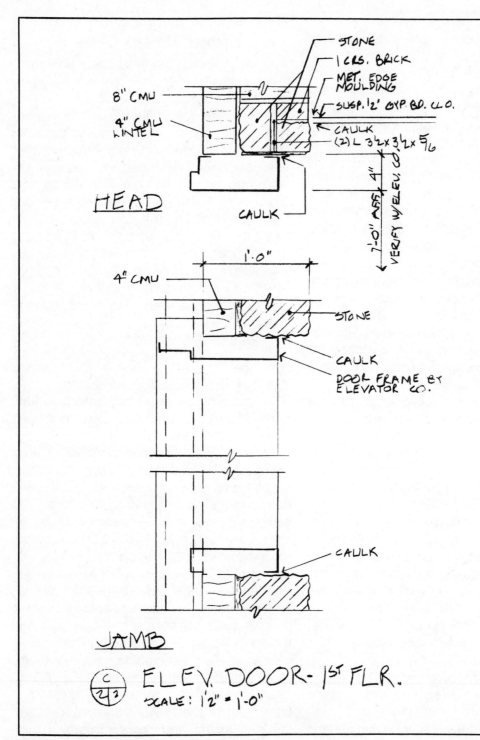

STONE

1 CRS. BRICK

MET. EDGE MOULDING

SUSP. $\frac{1}{2}'$ GYP. BD. CLG.

8" CMU

4" CMU LINTEL

CAULK

(2) L $3\frac{1}{2} \times 3\frac{1}{2} \times \frac{5}{16}$

HEAD

CAULK

7'-0" AFF VERIFY W/ELEV. CO.

4"

1'-0"

4" CMU

STONE

CAULK

DOOR FRAME BY ELEVATOR CO.

CAULK

JAMB

$\frac{C}{2\cdot 2}$ ELEV. DOOR - 1ST FLR.
SCALE: $1\frac{1}{2}" = 1'-0"$

Fig. 9-8. Sectional views of a commercial building.

132

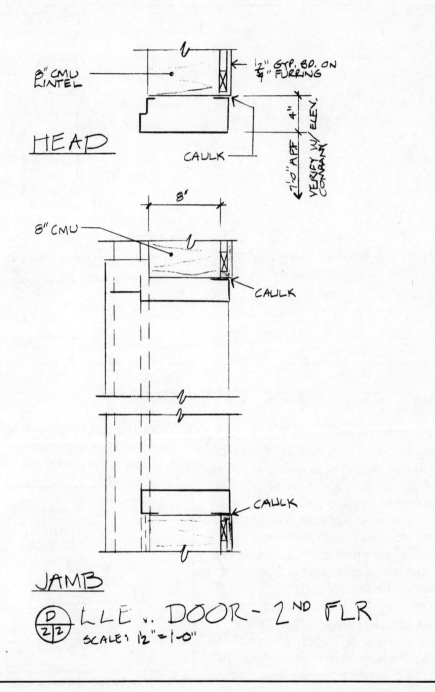

8" CMU
LINTEL

½" GYP. BD. ON
¾" FURRING

HEAD

CAULK

4"

1'-0" APP.

VERIFY W/ ELEV.
COMPANY

8'

8" CMU

CAULK

CAULK

JAMB

D / 2 / 2 LLE v. DOOR - 2ND FLR
SCALE: 1½" = 1'-0"

133

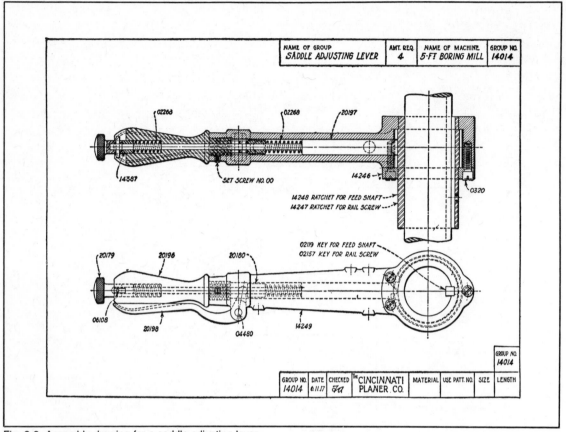

Fig. 9-9. Assembly drawing for a saddle adjusting lever.

READING WORKING DRAWINGS

When reading working drawings of any type, you will frequently run across simplified versions of an object—again, to shorten the time required to do the drawing. With experience, you will be able to look at a drawing of an object that you have never seen before and determine immediately what the drawing represents. Don't expect to become this proficient right away.

Some drawings may not have the conventional border line and are shown on separate sheets of paper. Instead, the drawings themselves will be mingled with paragraphs describing how an object is built, assembled, etc. This technique is often helpful when trying to learn something new in that a written description is given and a picture immediately follows the sentence.

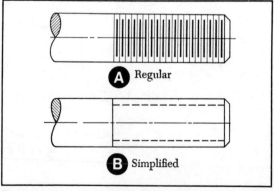

Fig. 9-10. Two methods of showing threads on machine shop drawings.

134

More often, however, the text will refer you to a nearby illustration. Working drawings are seldom done in this manner.

Most working drawings are made on drawing paper from 8½ × 11 inches in size up to 11 × 17 inches, 24 × 36 inches, or larger. Most have a border line drawn around the perimeter to enclose the required data. Each sheet will normally have a title block, and drawing sheets will be numbered consecutively to keep them in order. In reading the drawings you would normally start with page one and review each sheet in order. You will probably want to review some of the more complicated parts before attempting to construct the project. During construction you will also want to refer to various details frequently to ascertain correct measurements and other points of interest.

Chapter 10

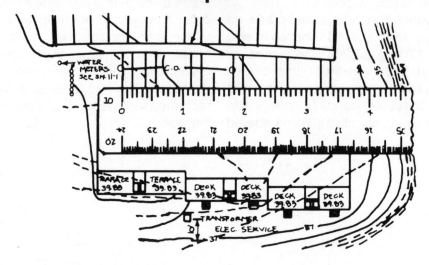

Assembly Drawings

ASSEMBLY DRAWINGS MAY RANGE FROM SIMPLE drawings on how to assemble a child's toy to those drawings accompanying a home computer kit. Assembly drawings usually deal with objects or components that have already been manufactured but have not yet been put in their proper order. The completion of the project sometimes may require that a hole be drilled and tapped or that an electrical connection be made, but usually the majority of the items have already been manufactured.

The average homeowner will view assembly drawings more than any other types, unless he or she happened to be an architect or construction engineer. Few items of any size come from the store fully assembled. A piece of furniture, an appliance, a child's toy, lawn equipment, and a shop tool all come from the manufacturer only partially assembled to make shipping easier. Local store owners may have one model assembled in the store so you can see what you are buying.

In a detail assembly drawing usually all the construction details and assembly of parts are shown on one drawing or a set of drawings. Note the assembly drawing of a chamber ironing tool in Fig. 10-1. All details are given as to material, dimensions, and shape of the object. Notes describing heat treatment and assembly details are also included. Any experienced machinist should be able to construct and assemble this tool without any difficulty.

Assembly drawings also include diagrams that clearly show the assembly or installation of an object or system in either pictorial or orthographic views. Such drawings are not made to any specific scale in most cases, and symbols are used to represent the various components. An electronic schematic diagram (Fig. 10-2) will show the various components by means of electronic symbols and their relationship to each other, along with connections, but these will be in diagrammatic form only. None are drawn to any particular scale. The only dimensions included are those that indicate distances between important points and are essential for assembly.

Display assembly drawings show the actual shape of each part of an object or assembly with all the pieces placed in their proper assembly position. Each part is identified, and descriptive notes are often added for clarity. These drawings are necessary for the average person who does not understand multiview drawings.

Some assembly drawings will be used to make several sizes of an object using only one drawing. A certain shape will be maintained, but the dimensions will be changed. This type of assembly drawing is called tabulated drawing. In this type of drawing a pictorial representation of a single item is shown, with variable dimensions coded by means of letters used as headings for columns in the tabulation. The variables are entered in the table under the appropriate headings and on the same line as the indentifying number or letter of the item on the drawing.

Assembly drawings use practically every type of drawing technique available to more clearly show how an object or system

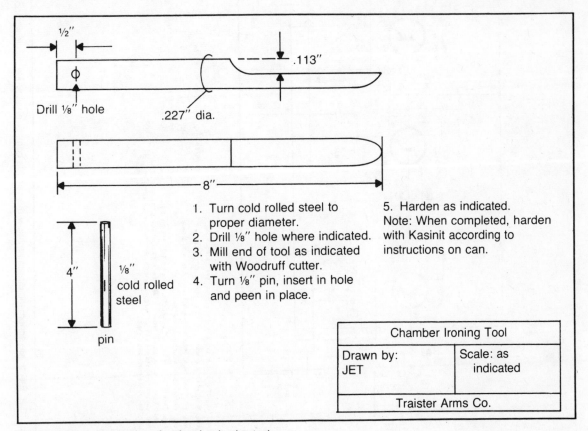

Fig. 10-1. Assembly drawing of a chamber ironing tool.

137

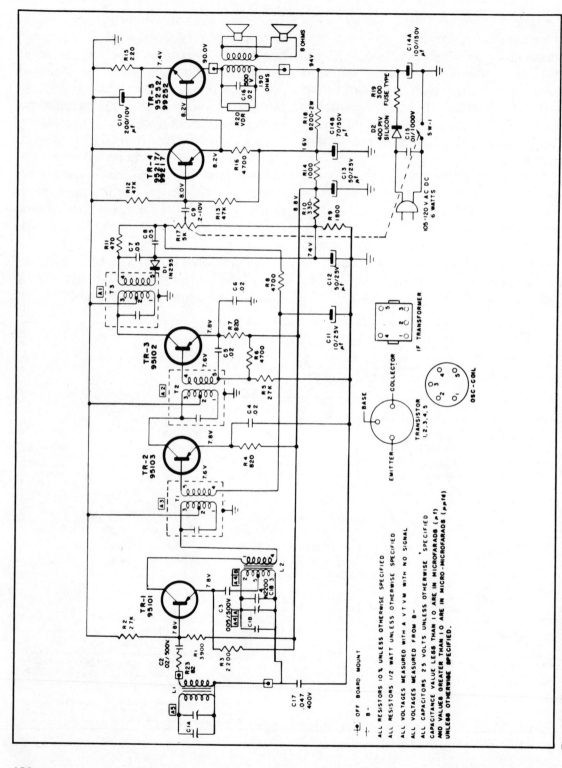

Fig. 10-2. Electronic schematic diagrams show the various components by electronic symbols.

138

Fig. 10-3. Step-by-step drawings showing the operation of a rifle sling.

IN OILY LOCATIONS   FOR OUTLETS THAT   OUTSIDE WIRING   CRAMPED AREAS   AROUND OBSTRUCTIONS   CONTOUR HUGGING
                           DON'T LINE UP

Fig. 10-4. Graphic representations of instructional material help the reader to remember the data longer.

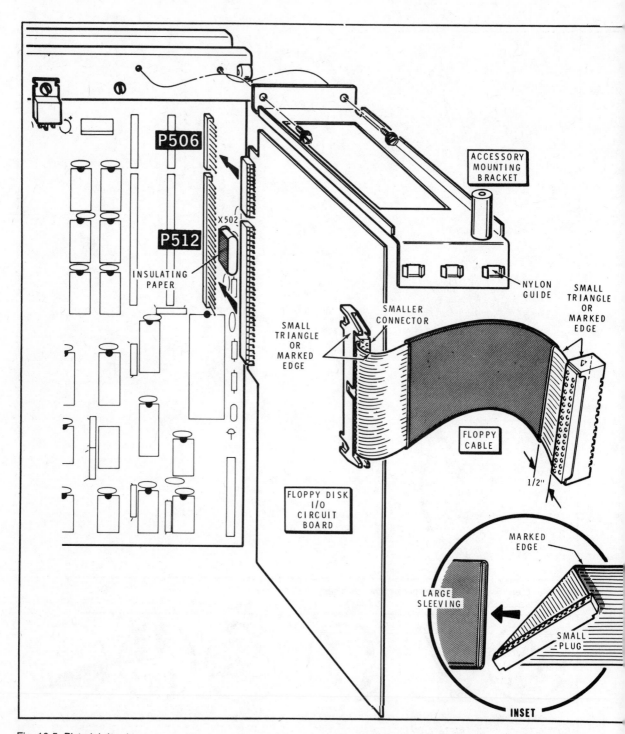

Fig. 10-5. Pictorial drawings are used in many electronic kits to make assembly easier (courtesy Heath Company, Benton Harbor, MI 49022).

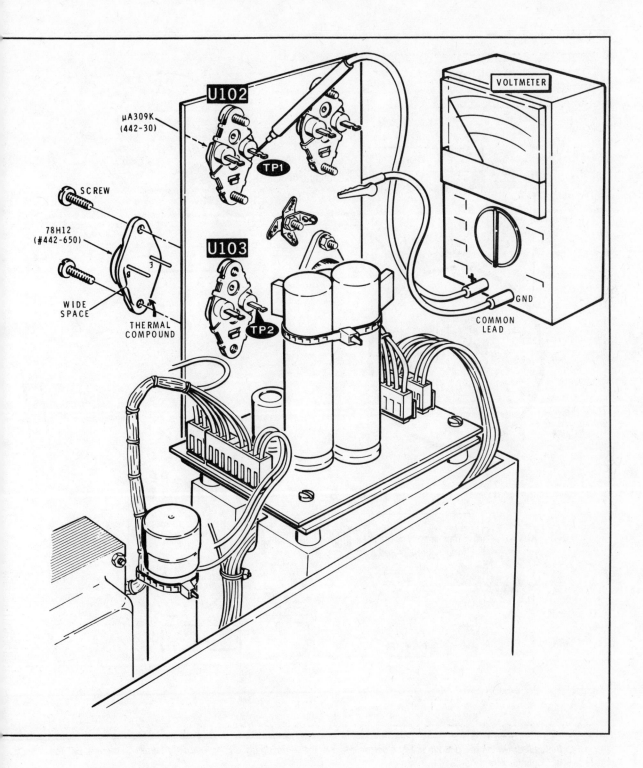

VOLTMETER

U102

μA309K
(442-30)

TP1

SCREW

78H12
(#442-650)

WIDE
SPACE

THERMAL
COMPOUND

U103

TP2

GND

COMMON
LEAD

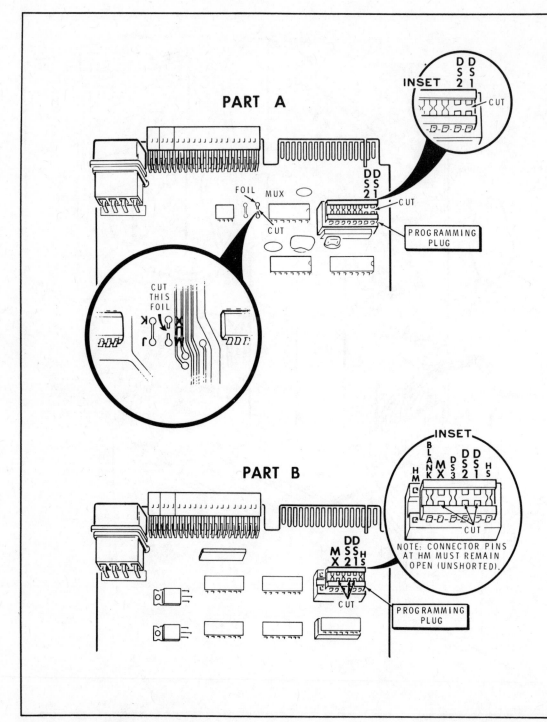

Fig. 10-5. Pictorial drawings are used in many electronic kits to make assembly easier (courtesy Heath Company, Benton Harbor, MI). Continued from page 141.

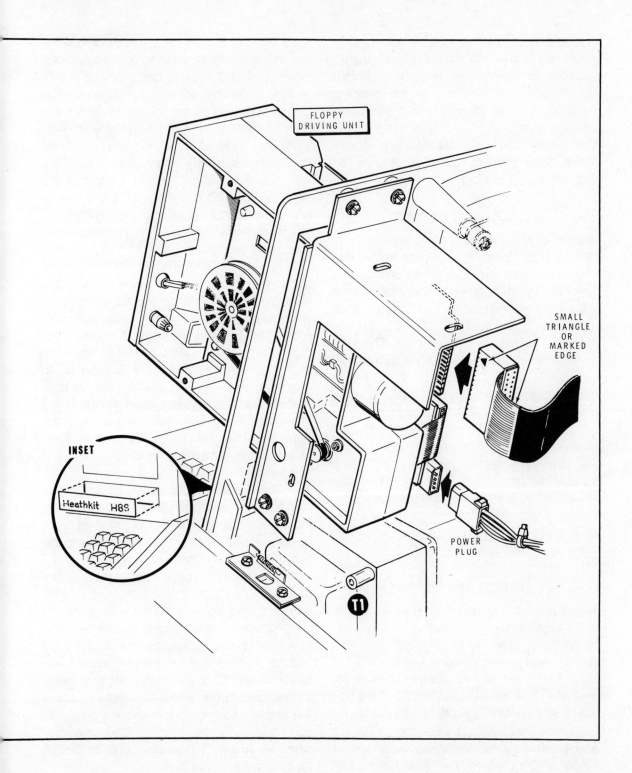

FLOPPY
DRIVING UNIT

SMALL
TRIANGLE
OR
MARKED
EDGE

INSET

Heathkit H89

POWER
PLUG

⑪

is assembled or installed. Orthographic views are common when the drawings are to be used by experienced workmen. Pictorial drawings are used whenever the drawings must be read by inexperienced personnel. Assembly drawings will also have schedules, sectional views, and auxiliary views to make each drawing as clear and understandable as possible.

INSTRUCTIONAL DRAWINGS

Instructional drawings come under the same category as assembly drawings in that they show how an object or system is operated. They are used mostly when written instructions may be difficult to understand. Most use notes or written instructions to supplement the drawings, or vice versa. The step-by-step drawings in Fig. 10-3 accompany a Latigo rifle sling and were provided by the manufacturer to simplify the instructions on how to adjust the sling.

Another graphic representation of instructional material is shown in Fig. 10-4. This drawing gives the general application of a particular brand of metal hose. The notes under each view conveys the same information as shown in the drawings, but with the addition of the drawings, the reader will remember information longer.

PRACTICAL APPLICATIONS

The pictorial drawings in Fig. 10-5 were used in an electronic kit supplied by Heath Company. They show very clearly just how certain phases of construction are carried out. Note also that intricate details are depicted perfectly. The insets, for example, show how the programming plug is to be installed in the unit, which connecting links are to be cut, and which are to remain intact. Drawings like these enable relatively inexperienced per-

sons to build do-it-yourself electronic kits.

An assembly drawing of the famous Savage Model 99 lever action rifle is shown in Fig. 10-6. This drawing enables owners, gunsmiths, and others to see each component used to complete the rifle and also the components' relationship to each other. This facilitates assembly and disassembly. The light solid lines show the exact position into which each part goes or fits. Each part is numbered to match an accompanying parts list (not shown here) so that replacement parts can be ordered.

While Fig. 10-6 should enable the experienced gunsmith to disassemble the weapon, step-by-step instructions should accompany the drawing for best results. For more complicated firearms, the manufacturer sometimes furnishes supplemental drawings such as the cross section of an action in Fig. 10-7—showing the various parts in their respective places. An exploded view, detail drawings, and written instructions should enable almost anyone to completely disassemble or disassemble a gun without damaging it.

The drawings discussed so far in this chapter have been somewhat technical. You may find that the drawings don't fit your needs, especially for applications around the home. There are drawings that show how to install a defective washer in a laundry sink, how to replace a plug fuse, how to hang curtains, and how to attach a cord on an electric iron. Once the job is finished, the instructions and drawings should be filed away in a safe place for later reference if you need replacement parts or want to change the location of the items. Let's look at some frequently used items around the home to see how supplemental assembly drawings can help you with the installation, repair, or maintenance of the items.

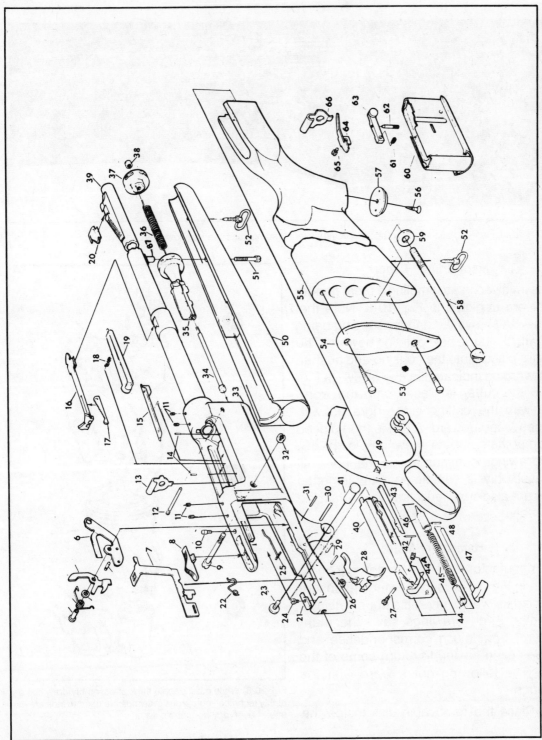

Fig. 10-6. Firearm manufacturers utilize assembly drawings to facilitate the assembly and disassembly of each weapon, along with providing a way of identifying the various parts.

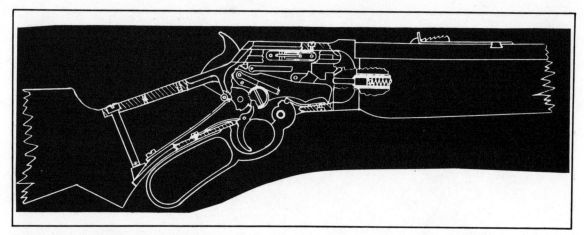

Fig. 10-7. To supplement assembly drawings, sectional views are sometimes used to show how certain mechanisms operate.

Meat Grinder Cutter

Steps involved in assembling a meat grinder cutter are depicted in Fig. 10-8. Note that Fig. 10-8A shows the grinder housing being held with the large hole pointing upward. The worm is placed into the housing and pushed downward as indicated by the arrow. In Fig. 10-8B the cutter is placed onto the worm shaft with the cutting side engraved with crosses faced upward. A grill is then placed on top of the cutter as shown in Fig. 10-8C. The drawings continue in this manner until the assembly is completed. Supplemental notes are also supplied to further explain the assembly.

Hamburger Mold

The manufacturer of the hamburger mold shown in Fig. 10-9 used drawings to ensure that the user would have little trouble using it correctly. Such drawings save the manufacturer from much correspondence and perhaps even having to return some of the units. The following notes were with the drawings.

●Place the removable disk inside the

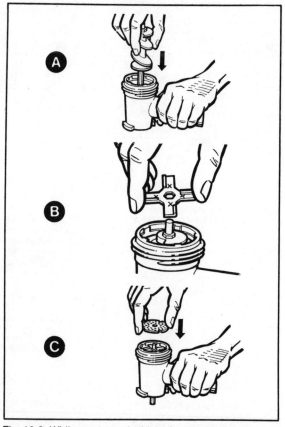

Fig. 10-8. While most people think of assembly drawings as highly technical, numerous drawings are used to indicate the use of everyday household items.

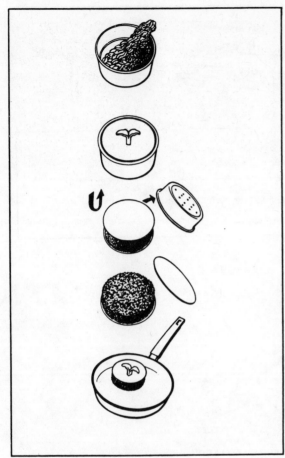

Fig. 10-9. Using a hamburger mold should be simple, but the manufacturer of the one shown here uses drawings to ensure that it will be used correctly.

mold. Put in the desired amount of minced meat.

●Press the meat down by means of the cover so that it takes the shape of the mold.

●Turn the mold upside down and remove the cover. The meat will remain stuck to the cover.

●Take away the removable disk.

●Hold the cover by its handle over the frying pan or grill, turn it over, and the meat will fall off and drop neatly into the cooking appliance.

Curtain Rod

Figure 10-10 will help with the installation of a curtain rod. The drawing of the assembled rod gives the viewer an overall picture of what the finished product will look like. The purpose of this drawing is to get the installer thinking correctly before he or she starts the assembly. The bracket location drawing shows how to take measurements as described in the printed instructions accompanying the curtain rod. The other details shown supplement the written instructions.

Fasteners

Many items used in the home must be secured to a wall or other surface. Because most homeowners install these items themselves, many manufacturers supply simple installation instructions to prevent the product or people from becoming damaged or hurt due to the object falling from the wall. Screws or nails are the most used fastening devices on wood, but some inexperienced home repairmen are not so knowledgeable about fastening to other materials such as gypsum board and paneling. Nor are they familiar with the names of the various fastening devices. Manufacturers found it necessary to picture the recommended fastening devices with supplemental notes. You may already be familiar with the fastening devices in Fig. 10-11, but let's review them.

Toggle Bolt. Drill an appropriate size hole in drywall partitions, masonry blocks, etc. The arms of the toggle bolt fold for insertion into the hole and then open and lock when they clear the further side of the drilled hole. Once in place, tighten the bolt with a screw slot.

Molley Screw Anchor. Drill a hole using

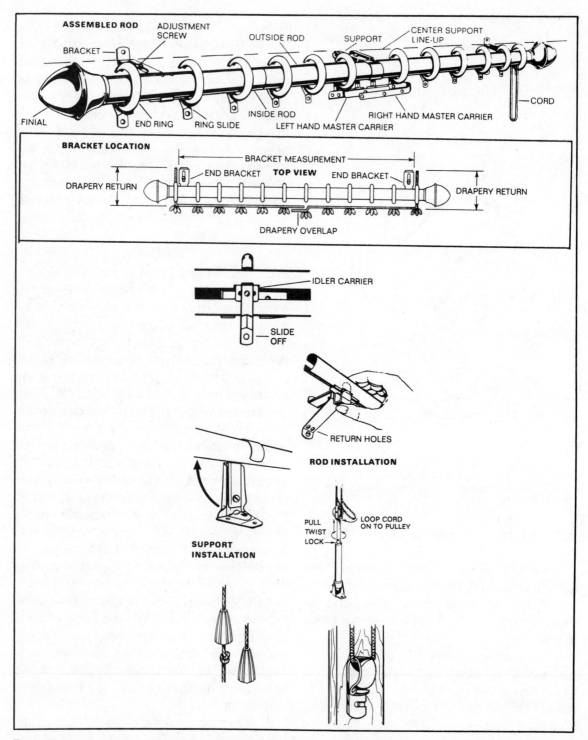

ASSEMBLED ROD

BRACKET
ADJUSTMENT SCREW
OUTSIDE ROD
SUPPORT
CENTER SUPPORT LINE-UP
CORD
FINIAL
END RING
RING SLIDE
INSIDE ROD
LEFT HAND MASTER CARRIER
RIGHT HAND MASTER CARRIER

BRACKET LOCATION

BRACKET MEASUREMENT
END BRACKET
TOP VIEW
END BRACKET
DRAPERY RETURN
DRAPERY RETURN
DRAPERY OVERLAP

IDLER CARRIER
SLIDE OFF

RETURN HOLES
ROD INSTALLATION

SUPPORT INSTALLATION

PULL
TWIST
LOCK
LOOP CORD ON TO PULLEY

Fig. 10-10. An assembly drawing of a curtain rod is another example of how assembly drawings are useful around the home.

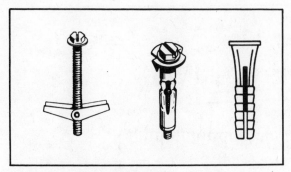
Fig. 10-11. Graphic representations of nearly every item imaginable are currently supplied with written descriptions.

the proper size drill bit. Tap the anchor into the hole and turn the screw until tight. Once the anchor is secure, the screw may be removed for the mounting bracket and then reinserted and tightened.

Plastic Screw Anchor. This anchor is used in plaster, wallboard, and similar materials. Drill a hole slightly smaller than the plug. Tape in the plug until flush with the wall, then insert screws.

The previous example was taken from installation instructions for an item meant to be secured to a wall. The manufacturers of the fastening devices may also include drawings with the fasteners to further help with proper installation. Photos are sometimes used to show step-by-step procedures, but drawings make it possible to see all the way around an object and are usually preferred over photos. Drawings are usually more costly to supply than photos.

Furniture

Probably one of the best examples of assembly drawings is the type used to show how to build a piece of furniture like a table, chair, or workbench. Home project books and magazines are full of these drawings.

Most readers would find it impossible to build the items described without the drawings.

Instructions for a typical how-to project will usually include a drawing (or photograph) of the finished item to show the builder what the finished product will look like. This is usually followed by a bill of materials required to complete the project—often keyed to the drawings by means of letters or numbers. Orthographic views will show plan, elevation, side, and any other views necessary to complete the project. Most magazines are now using exploded views to show the relationship of each part to the others. You will find details of joints, moldings, and similar items requiring a better explanation to construct the project correctly. Notes appearing on the drawings, dimensions, and step-by-step instructions complete the set of construction plans.

CONSULTING MAGAZINES AND BOOKS

Look through several magazines containing how-to projects such as *Popular Science, Mechanix Illustrated, Workbench*, etc., and study the drawings contained in each. Note all dimensions and how they were applied to the drawing. Determine why certain portions of the project are better explained than others. Read over the building instructions and see if you can construct the project in your mind. Continue with this procedure until understanding of the projects becomes second nature. You will then have gained a good knowledge of reading assembly drawings.

There are also many excellent books on how-to projects that contain drawings for study purposes. Check your local library or see the list offered by TAB Books, Inc.

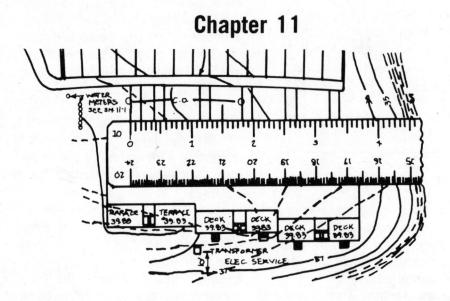

Chapter 11

Electrical and Electronic Drawings

ELECTRICAL AND ELECTRONIC DRAWINGS PREpared by consulting engineers, contractors, manufacturers, and others are very unique. Most of these drawings will encompass all of the previously described drawings regardless of the firm involved. Drawings produced by manufacturers of electronic equipment (Radio Shack, Heath Company, etc.) usually consist of the following:

●A schematic diagram (Fig. 11-1) to show—in diagrammatic form—the connections of the various components for the project.

●Pictorial drawings to illustrate the various components (Fig. 11-2) or to show more clearly how the mechanical and electrical work is to be done on a project (Fig. 11-3). Perspective drawings will also be required to illustrate the piece of equipment in catalogs.

●Orthographic views to show how such items as, say, an equipment chassis are to be constructed.

Electric lighting manufacturers need orthographic views as a guide for manufacturing lighting fixtures and for use as reference by consulting engineers and electrical contractors. See Fig. 11-4. They may also require oblique or isometric drawings similar to the one in Fig. 11-5.

Of the types of drawings previously mentioned, the diagrams are probably the type most used in all electrical and electronic drawings. Many electrical and electronic drawings deal with circuits and their components.

SCHEMATIC WIRING DIAGRAMS

Schematic wiring diagrams represent components in the system by symbols (see Chapter 2), the wiring and connection to each, and other detailed information. Sometimes the wires are shown in an assembly of several wires, which appear as one line on the drawing. When this method is used, each wire should be numbered where it enters the assembly and should keep the same number when it comes out of the assembly to be connected to some component in the system. If the schematic does not follow this procedure, mark and number the wires yourself.

Although the symbols represent certain components, an exact description of each is usually listed in schedules or else noted on the drawings. Such drawings are seldom, if ever, drawn to scale as an architectural or cabinet drawing would be. They appear in diagrammatic form. In the better drawings, however, the components are arranged in a neat and logical sequence so that they are easily traced and can be understood easily.

Whereas schematic drawings are the type most frequently used on electronic drawings, block diagrams—known as power-riser diagrams—are the type most often drawn on electrical working drawings for building construction (Fig. 11-6). These diagrams indicate what components are to be used and how they are to be connected in relation to one another. This type of diagram is easily understood and requires much less time to draw or interpret than a schematic diagram—either detailed or one-line.

If you are involved in electronic projects of any nature, you must be able to interpret electronic drawings and to understand the theory behind the drawings. A good understanding will only come with experience and by studying other TAB books on electronics.

One of the best ways to begin is by doing—by actually building electronic projects. There are many useful electronic gadgets available in kit form from Radio Shack, Heath Co., etc. These kits come with easy-to-follow directions, and most contain a pictorial drawing and a schematic drawing that you follow while constructing the kits. Refer to the pictorial drawing and then compare this with the schematic diagram. Start off with an easy kit (costing very little) and work up to the more sophisticated models. There is probably no better way to quickly learn how to read electronic schematic diagrams.

Electronic schematic diagrams indicate the scheme or plan according to which electronic components are connected for a specific purpose. Diagrams are not normally drawn to scale, and the symbols rarely look exactly like the component. Lines joining the symbols representing electronic components indicate that the components are connected.

To serve all its intended purposes, the schematic diagram must be accurate. Also, it must be understood by all qualified personnel, and it must provide definite information without ambiguity.

The schematics for an electronic device should indicate all circuits in the device. If they are accurate and well prepared, it will be easy to read and follow an entire closed path in each circuit. If there are interconnections, they will be clearly indicated.

In nearly all cases the conductors connecting the electronic symbols will be drawn either horizontally or vertically. Rarely are they ever slanted.

A dot at the junction of two crossing wires means a connection between the two wires. An absence of a dot in most cases indicates that the wires cross without connecting.

Schematic diagrams are, in effect,

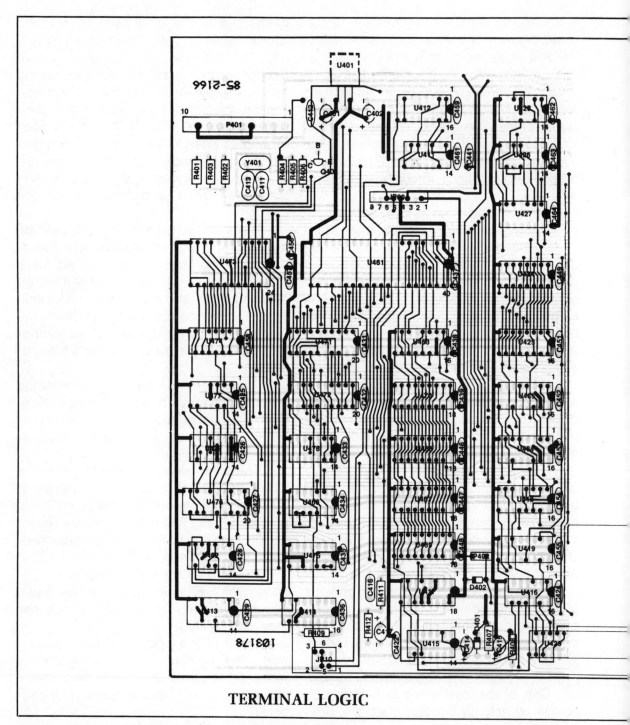

TERMINAL LOGIC

Fig. 11-1. Schematic diagrams are frequently used on electronic projects to show the location of components and the connections between each (courtesy Heath Company, Benton Harbor, MI 49022).

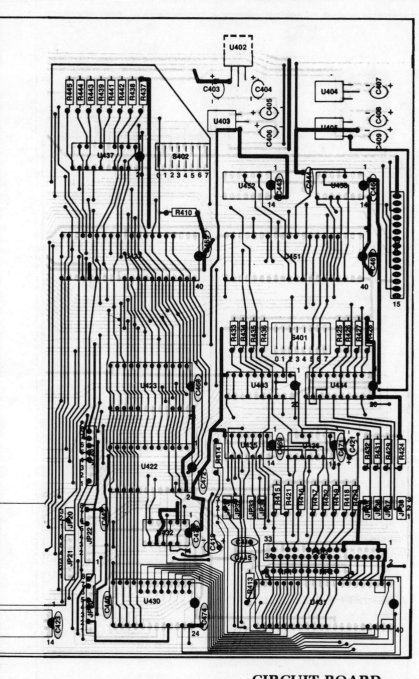

CIRCUIT BOARD

153

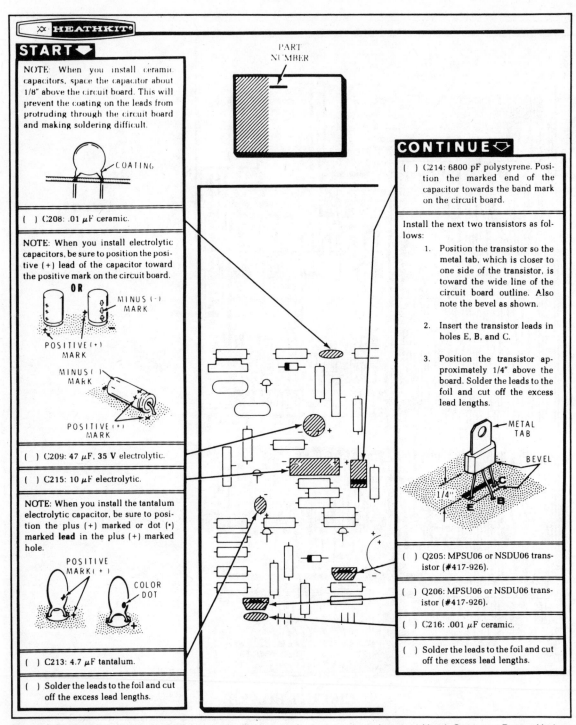

HEATHKIT

START

NOTE: When you install ceramic capacitors, space the capacitor about 1/8" above the circuit board. This will prevent the coating on the leads from protruding through the circuit board and making soldering difficult.

COATING

( ) C208: .01 μF ceramic.

NOTE: When you install electrolytic capacitors, be sure to position the positive (+) lead of the capacitor toward the positive mark on the circuit board.

OR

MINUS (-) MARK

POSITIVE (+) MARK

MINUS (-) MARK

POSITIVE (+) MARK

( ) C209: 47 μF, 35 V electrolytic.

( ) C215: 10 μF electrolytic.

NOTE: When you install the tantalum electrolytic capacitor, be sure to position the plus (+) marked or dot (•) marked lead in the plus (+) marked hole.

POSITIVE MARK (+)

COLOR DOT

( ) C213: 4.7 μF tantalum.

( ) Solder the leads to the foil and cut off the excess lead lengths.

PART NUMBER

CONTINUE

( ) C214: 6800 pF polystyrene. Position the marked end of the capacitor towards the band mark on the circuit board.

Install the next two transistors as follows:

1. Position the transistor so the metal tab, which is closer to one side of the transistor, is toward the wide line of the circuit board outline. Also note the bevel as shown.

2. Insert the transistor leads in holes E, B, and C.

3. Position the transistor approximately 1/4" above the board. Solder the leads to the foil and cut off the excess lead lengths.

METAL TAB

BEVEL

1/4"

E B C

( ) Q205: MPSU06 or NSDU06 transistor (#417-926).

( ) Q206: MPSU06 or NSDU06 transistor (#417-926).

( ) C216: .001 μF ceramic.

( ) Solder the leads to the foil and cut off the excess lead lengths.

Fig. 11-2. Pictorial drawings are sometimes used to illustrate various components (courtesy Heath Company, Benton Harbor, MI 49022).

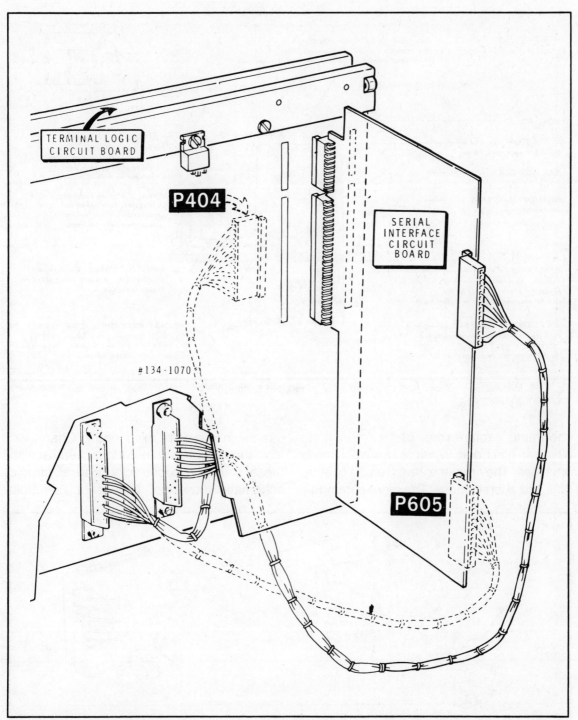

TERMINAL LOGIC
CIRCUIT BOARD

P404

SERIAL
INTERFACE
CIRCUIT
BOARD

#134-1070

P605

Fig. 11-3. Pictorial drawings are also used to more clearly show how various components are arranged (courtesy Heath Company, Benton Harbor, MI 49022).

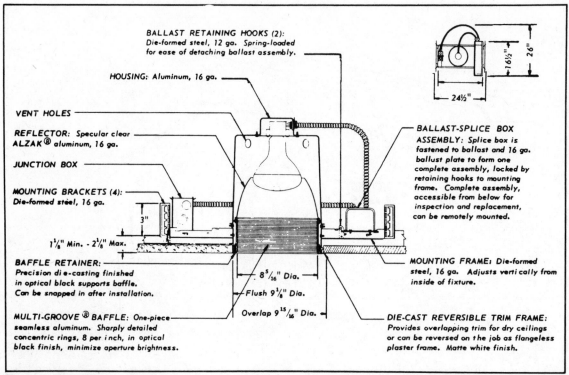

BALLAST RETAINING HOOKS (2): Die-formed steel, 12 ga. Spring-loaded for ease of detaching ballast assembly.

HOUSING: Aluminum, 16 ga.

VENT HOLES

REFLECTOR: Specular clear ALZAK® aluminum, 16 ga.

JUNCTION BOX

MOUNTING BRACKETS (4): Die-formed steel, 16 ga.

$1\frac{1}{8}$" Min. - $2\frac{1}{8}$" Max.

BAFFLE RETAINER: Precision die-casting finished in optical black supports baffle. Can be snapped in after installation.

MULTI-GROOVE® BAFFLE: One-piece seamless aluminum. Sharply detailed concentric rings, 8 per inch, in optical black finish, minimize aperture brightness.

3"

$8\frac{5}{16}$" Dia.

Flush $9\frac{1}{8}$" Dia.

Overlap $9\frac{15}{16}$" Dia.

BALLAST-SPLICE BOX ASSEMBLY: Splice box is fastened to ballast and 16 ga. ballast plate to form one complete assembly, locked by retaining hooks to mounting frame. Complete assembly, accessible from below for inspection and replacement, can be remotely mounted.

MOUNTING FRAME: Die-formed steel, 16 ga. Adjusts vertically from inside of fixture.

DIE-CAST REVERSIBLE TRIM FRAME: Provides overlapping trim for dry ceilings or can be reversed on the job as flangeless plaster frame. Matte white finish.

$16\frac{1}{2}$" 26"

$24\frac{1}{2}$"

Fig. 11-4. Manufacturers of electrical equipment may be required to furnish orthographic views for use as reference by engineers and contractors.

shorthand explanations of the manner in which an electronic circuit or group of circuits operates. They make extensive use of symbols and abbreviations. The more commonly used symbols were explained in Chapter 2. You should become thoroughly familiar with these symbols to correctly read electronic schematic drawings. The use of symbols

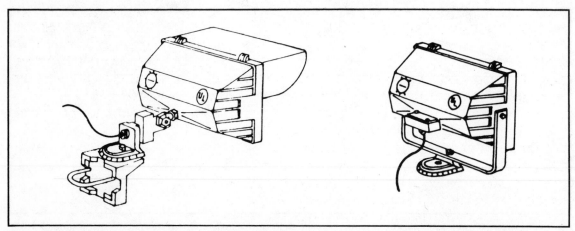

Fig. 11-5. Oblique or isometric drawings are sometimes used by manufacturers to clearly illustrate their product.

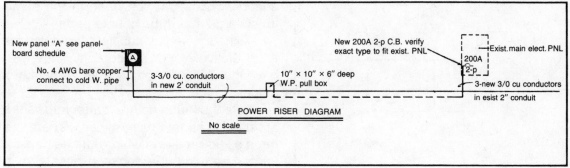

Fig. 11-6. Power-riser diagrams may be classified as block drawings.

presumes that the person looking at the diagram is reasonably familiar with the operation of the device, and that he will be able to assign the correct meaning to the symbols. If the symbols are unusual, a legend will normally be provided to clarify matters.

Every component on a complete schematic diagram usually has a number to identify the component. Supplementary data about such parts are supplied on the diagram or on an accompanying list in the form of a schedule (see Chapter 3), which describes the component in detail or refers to a common catalog number familiar in the trade.

To interpret schematic diagrams, remember that each circuit must be closed in itself. Each component should be in a closed loop connected by conductors to a source of electric current such as a battery, transformer, or other power supply. There will always be a conducting path leading from the source to the component and a return path leading from the component to the source. The path may consist of one or more conductors. Other components may also be in the same loop or in additional loops branching off to other devices. For each electronic component, it must be possible to trace a completed conducting loop to the source.

IDENTIFYING RESISTORS

In schematic drawings resistors are shown by the symbol with an identifying note such as R_1, R_2, etc., which gives the value in an accompanying schedule; or the type and value of the resistors are called out directly on the drawing. A symbol for a resistor may have the value .01 μF lettered beside it. You may have to identify the various resistors by color codes in pictorial assembly drawings.

Most ½ through 2-watt resistors are identified by colored bands on the resistor. Any person working with electronic projects should be able to identify the value of these resistors.

To read the color code, the resistor should be viewed with the colored bands towards the left like Fig. 11-7. The colored bands are nearer to one end than the other on the resistor in Fig. 11-7. The colored bands may be centered on other resistors. The gold or silver band should be on the right-hand side as it is viewed.

Each color represents a number as follows:

Black	0
Brown	1
Red	2
Orange	3
Yellow	4

157

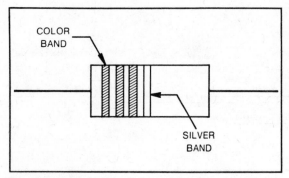

Fig. 11-7. Method of identifying certain types of resistors.

Green	5
Blue	6
Violet	7
Gray	8
White	9

A resistor with the colors (from left to right) yellow (4), violet (7), and brown (1) is read as follows. The first yellow band represents the number 4, and the second band represents the number 7. The two combined colors gives a value of 47. The third color (brown) gives the number of zeros in the value. Because brown represents 1, the third band indicates that the resistor has one zero in its value. The value of the resistor is 470 ohms.

If the resistor had been color-coded yellow, violet, and red, the first two figures would again have the value of 47 because yellow = 4 and violet = 7, or 47. Red represents 2, so there would be two zeros. The value would be 4700 ohms—often written 4.7 k ohms, and the "k" stands for 1000. If the colors were yellow, violet, and orange, the value would be 47,000 ohms or 47 k ohms, etc.

SEMICONDUCTORS

People who worked on electronic projects of the past dealt with electron tubes. While some tubes are still in use, most electronic technicians and hobbyists deal mostly with semiconductors, which perform the same function as their tube counterparts.

Although semiconductors are noted for their relatively poor conductivity, they also have the ability to conduct electricity in two modes.

Like their tube counterparts, transistors are used in certain basic circuits. There are three basic circuits in which a triode transistor can be used: the common-base circuit, the common-emitter circuit, and the common-collector circuit. Of these three, the common-emitter circuit is found more frequently than the other two, mainly because this circuit gives the greatest voltage and power gain. The fundamentals of all three circuits will be briefly described to give you a basic knowledge of transistors.

Common-Base Circuit

A typical common-base circuit is shown in Fig. 11-8. This is for an npn transistor. The solid arrows indicate the direction of useful electron flow. Electrons flow from the negative terminal of battery B1, through resistor R1, through the npn transistor, into the emitter, across the emitter-base junction to the base, then across the base-collector junction into the collector, and from the col-

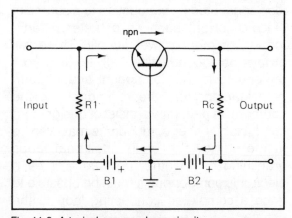

Fig. 11-8. A typical common-base circuit.

lector through the collector resistor R2 to the positive terminal of battery B2.

Characteristics of the common-base transistor circuit include a very low input impedance and a high output impedance. This output voltage is in phase with the input voltage and is greater than the input voltage. The current gain of the stage is less than one. A common-base circuit is often used in applications for matching a low impedance to a high impedance.

Common-Emitter Circuit

The most frequently used transistor circuit is the common-emitter circuit shown in Fig. 11-9. The circuit shown utilizes an npn transistor. The battery B1 provides the forward bias needed for the emitter-base junction, and battery B2 provides the reverse bias needed for the base-collector junction.

The emitter is made negative with respect to the base to provide forward bias for the emitter-base junction of the npn transistor. The solid arrows in Fig. 11-9 indicate the direction of electron flow in the circuit. Electrons leave the negative terminal of battery B1 and flow to the emitter. They cross the emitter-base junction, then flow through the base, across the base-collector junction,

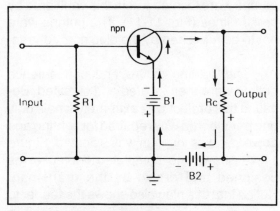

Fig. 11-9. A common-emitter circuit.

and through resistor R2 to the positive terminal of battery B2.

As stated previously, the common-emitter circuit is the most frequently used transistor circuit and has a voltage gain of from 80 to 100. There is also considerable current gain in this type of circuit. It has a medium input resistance and an output resistance of about 20,000 ohms. The output signal is 180 degrees out of phase with the input signal.

Common-Collector Circuit

The third possible circuit configuration using a triode transistor is the common-collector circuit (Fig. 11-10). In this type of circuit the collector is operated at signal ground potential. Although this circuit is not found as often as the common-emitter circuit, it does have characteristics that are useful in special applications.

Other types of semiconductors include diodes that are used in rectifying applications. Their inherent characteristics permit many uses other than simply as signal and power rectifiers. Table 11-1 lists special diode applications.

ELECTRICAL DRAWINGS

Electrical drawings must be read in every branch of electrical work. Electricians responsible for installing the electrical system in a new building usually consult an electrical drawing to locate the various outlets, the routing of circuits, the location and size of panelboards, and other similar electrical details in preparing a bid. The electrical estimator of a contracting firm must refer to electrical drawings to determine the quantity of material needed. Electricians in industrial plants consult schematic diagrams when wiring electrical controls for machinery.

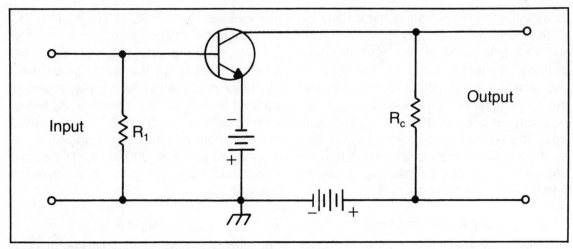

Fig. 11-10. A triode transistor is used in the common-collector circuit.

Plant maintenance men use electrical blue-prints in troubleshooting. Circuits may be tested and checked against the original drawings to help locate any faulty points in the installation.

The ideal electrical drawing should show in a clear, concise manner exactly what is required of the workmen. The amount of data shown on such a drawing should be sufficient but not overdone. A complete set of electrical drawings could consist of one sheet of paper or several dozen sheets, depending on the complexity of the project.

To better understand the "reading" of an electrical drawing, let's analyze a small project and the thought behind the design.

An electrical design firm has been commissioned to provide the working drawings for a water-pumping station to demonstrate the project in question. The electrical designer was provided with the pertinent information—floor plan, elevation views of the pumping station vault, the location of two 10-horsepower (hp) pumps, and design criteria stating that an electrical service was required along with lighting and power outlets.

A sheet of tracing paper 24 inches by 36 inches was chosen to fit the owner's standard working-drawing sets. The electrical draftsman began the working drawings by placing heavy border lines ½ inch inside the outside perimeter of the drawing sheet. An outline for a title block was also drawn to be filled in later.

After carefully studying all the information, the draftsman finally decided on a floor-plan view of the vault showing the location of the wall partitions and door. The two pumps were located within this vault area by broken lines (Fig. 11-11). The pumps were not to be furnished by the electrical contractor.

The lighting fixture and convenience outlets were then located and circuited. Because this project was extremely small, only one panelboard was required for lighting and power, and its location was selected. Figure 11-12 shows the completed floor plan as designed and drawn by the draftsman. Notice that this plan also shows the feeder to

Table 11-1. Semiconductor Devices and Applications.

Device	Junction Configuration or Type	Semiconductor Material	Application
Rectifiers (diodes)	pn junction	Silicon	Low forward voltage drop, signal, and power rectification.
	pn junction	Germanium	High rectification ratio, high inverse breakdown, high temperature, signal and power rectification.
	Dry disk	Selenium	Power rectifier, low-frequency diode, self-healing.
	Dry Disc	Copper oxide	Meter rectifier, low voltage power rectifier.
	Dry Disc	Copper sulfide	Low voltage power rectifier.
Transistors	pnp or npn	Germanium	Low saturation voltage, general purpose to 75 degrees Celsius.
	pnp or npn	Silicon	High temp. use to 175° Celsius higher voltage.
Field Effect Transistors (FET)	n or p channel types	Silicon	High input impedance, resistive, bidirectional output impedance.
Unijunction Transistors	n-type bar with p-junction between ends of bar	Silicon	Relaxation oscillator, timing, trigger circuits, negative resistive device.
Thyristors Silicon Controlled Rectifier (SCR)	pnpn	Silicon	Phase controlled rectifier, similar to thyratron tube.
Triac	two SCRs in parallel and opposite orientation	Silicon	Bidirectional control of ac, light dimmers, power tool speed control.
Tunnel Diode	heavily doped pn junction	Germanium Gallium arsenide	Negative resistance, microwave amplifier oscillators, converters.
Photodiodes	pn	Germanium	Photo-conduction and photovoltaic for control applications.
Phototransistors	npn	Silicon	
Photoelectric Cells	photoresistive	Lead sulfide	Infrared detector.
	photoresistive	Lead telluride	Infrared detector.
	photovoltaic	Cadmium sulfide	Light meter.
	photovoltaic	Selenium	Light meter.
Varistors	Fired	Silicon carbide	Surge suppressors.
	Dry disc	Selenium	Contact protector.
Thermistors	Fired	Mixed metal oxides	Temperature sensing, control, compensating.
Zener Diodes	pn junction reverse biased to zener breakdown	Silicon	Reverse biased, voltage regulator, voltage reference.
Varactor Diode (varicap)	pn junction reverse biased, no current flow	Silicon	Voltage controlled capacitor, parametric amplifier. multipliers.
Light Emitting Diode (LED)	pn junction	Gallium arsenide Phosphide	Visible, infrared, light emitting. displays.

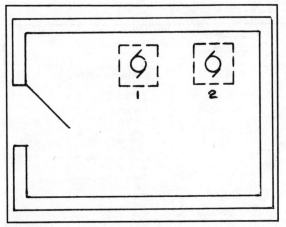

Fig. 11-11. Floor plan of a pumping station showing location of pumps.

each pump. The pumps chosen for this project were shipped from the manufacturer as a package with all of the controls built in, and thus no additional wiring details need be shown.

Table 11-2 shows a panelboard sched-ule that describes the panelboard components. A power-rise diagram (Fig. 11-13) shows the details of the service entrance. Notes lettered on the drawing give some extra data on the type of lighting fixtures required, etc. The completed electrical drawing is shown in Fig. 11-14.

When a building of any size is planned, an architect is usually commissioned to plan and design it. An engineer or electrical designer usually lays out the entire electrical system of the building for the architect. An electrical draftsman transforms the engineer's designs into complete working drawings. The following list shows the steps involved in preparing an electrical design and working drawings:

●The engineer meets with the architect and owner to discuss the electrical needs of the building and the recommendations made by all parties.

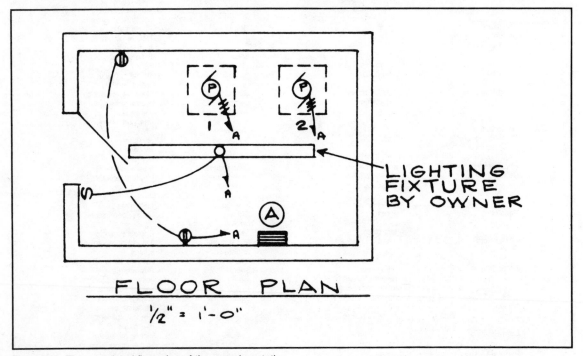

Fig. 11-12. The completed floor plan of the pumping station.

Table 11-2. Panelboard Schedule Used for the Pumping Station.

PNL. N°	TYPE CABINET	MAINS			BRANCHES			REMARKS
		AMPS	VOLT	PHASE	1-P	3-P	PROT	
A	SURFACE	100	120/240	3P4W	1	–	20	LTS.
SQ. 'D' TYPE NQO W/ 100-A. MAIN					1	–	20	RECEPTS
						2	30	PUMPS

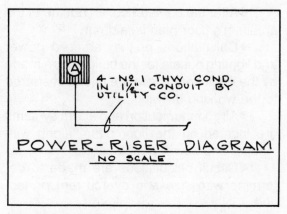

Fig. 11-13. Power-riser diagram used on the pumping station drawings.

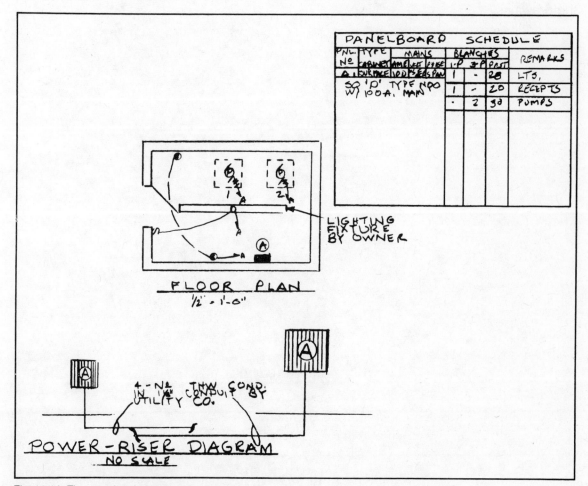

Fig. 11-14. The completed electrical drawing.

●After the conference, an outline of the architect's floor plan is laid out.

●Calculations of the required power and lighting outlets for the building are made by the engineer. These are later transferred to the working drawing.

●All communication and alarm systems are located on the floor plans along with lighting and power panelboards.

●Circuit calculations are made to determine wire size and overcurrent protection.

●The main electric service and related components are determined and shown on the drawings.

●Schedules are placed on the drawings to identify electrical equipment.

●Wiring diagrams are made to show workmen how various electrical components are connected.

●An electrical symbol list or legend is shown on the drawings to identify all symbols used for electrical outlets or equipment.

●To better help the workmen know what is expected, various large-scale electrical details are included.

●Written specifications are then made to describe the materials and installation methods.

Chapter 12

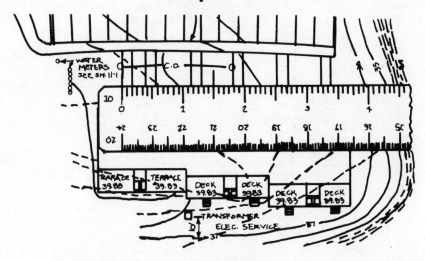

Mechanical Drawings

MECHANICAL DRAWINGS ARE BASCIALLY ACCU-rate, neat, clear line representations of mechanical objects drawn solely for the purpose of enabling workmen to make these objects. Every machine from a pencil sharpener to a guided missile, every machine part, every tool, every invention, and every building from a roadside fruit stand to a large skyscraper usually originates on the drawing board. Practically every man-made object, whether it's a screwdriver or a highly complex computer, is developed from mechanical drawings.

Mechanical drawings have many branches. The machinist sees mechanical drawings of cams, levers, nuts, bolts, and other parts of complicated machinery. The structural worker's drawings are confined to such items as I beams, columns, plate girders, rivets, and reinforced concrete design.

Mechanical drawings for building con-plumbing, heating, ventilating, air conditioning, and temperature control systems. Most of these drawings are highly diagrammatical and are used to locate pipes, fixtures, ductwork, equipment, etc.

The exact method of showing mechanical layouts on drawings for building construction will vary with the engineer and consulting firm. The following description is typical for most mechanical drawings.

PLUMBING DRAWINGS

Plumbing drawings cover the installation of the plumbing system within a building and on the premises. They cover the complete design and layout of the plumbing system and show floor-plan layouts, cross sections of the building, and detailed drawings.

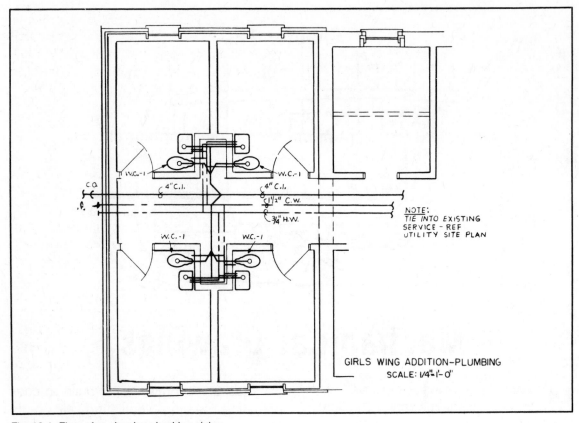

Fig. 12-1. Floor plan showing plumbing piping.

All domestic cold water piping is normally laid out on the building floor plans as shown in Fig. 12-1. Note that all valves, stops, and other connections are indicated by symbols. The water lines are indicated by broken lines. The ————-——— pattern on the drawing indicates a cold water line, and the ———--——— pattern indicates a hot water line. Always check the legend or symbol list on each set of plumbing drawings, as these symbols could be reversed or altogether different.

Drains and vents in the drawing in Fig. 12-1 are shown by symbols only; no piping is shown in the plan view. When such a case exists, a piping riser diagram such as the one shown in Fig. 12-2 is recommended. It diagrammatically shows the arrangement of the drain and vent pipes.

Pipe sizes and other pertinent details are usually indicated by numerals or notes respectively and appear immediately adjacent to the component or equipment described. Schedules and written specifications complete the set of plumbing working drawings. The schedules include the manufacturer's catalog number, pipe connection, etc., of all plumbing fixtures and similar equipment.

Let's look at a set of plumbing drawings used on a local project. By following through the drawing with the text, you should quickly become accustomed to reading plumbing drawings. See Fig. 12-3.

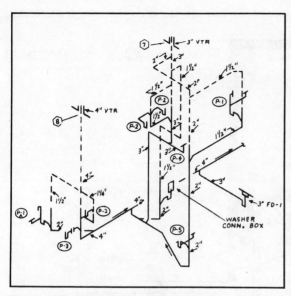

Fig. 12-2. Riser diagram of a plumbing drainage system.

Note the plan title at the very bottom of Fig. 12-3 that states, "1st Floor Plumbing Plan." Directly under this is the scale to which the drawing is made—⅛ inches = 1 foot-0 inch—meaning that ⅛ inch on the drawing represents 1 foot on the actual building.

The plan of the building in question is made as though the building was cut at the window line and the top half was lifted off to reveal the interior partitions, windows, and the like. On this drawing the plain partitions represent the existing building. The shaded partition represents a new addition to be built.

Before continuing, look at the abbreviations and symbol legend in Fig. 12-4. In looking over Fig. 12-3, continually refer to the legend for symbols you don't understand. Note that the existing waste lines are indicated by a solid *light* line. The new waste lines are shown by a *heavy* solid line. The arrowheads indicate the direction of sewage flow. Find these lines on Fig. 12-3.

You probably had little trouble finding the new waste pipes on the plumbing floor plan. It is probably unclear just what these waste pipes do, so look at the plumbing fixture schedule in Table 12-1. If you follow the new waste lines carefully on the drawing, you will find that the drain or waste line connects to "P-3" and FD-1. Checking Table 12-1 you will see that these two items are a sink and floor drain, respectively.

How about the other items in Table 12-1. Look at Fig. 12-5. As the title indicates, this is the second floor plumbing plan and is directly above the first floor plan. Note the two toilets and also the fixtures contained therein. Referring to Table 12-1 you will find that P1 is a water closet (W.C.), P-2 is a lavatory, and P-4 is a different lavatory. You will also find CO-1 (cleanout) in the upper right-hand corner of the right-hand toilet. Cleanout number 2 (CO-2) is located on the first floor plan (Fig. 12-3); see if you can find it.

If you refer again to Fig. 12-4, you will find other symbols that have yet to be mentioned. These include the hot and cold water lines, pipes turned up and down, and others. The symbols for pipe turned down and connection new to existing are not shown on this plan, but see if you can locate the other symbols in Figs. 12-3 and 12-5.

The notes shown in Table 12-2 were also included with this plumbing drawing to facilitate construction of the project. These drawings shown along with the written specifications (not shown here) complete the set of plumbing drawings.

On a larger project such as an industrial chemical plant, the plumbing and piping drawings will be quite extensive. Several sheets of drawings containing plain views, elevations, sections, and detail drawings along with piping riser diagrams are needed. Drawings usually utilize single lines for pipes

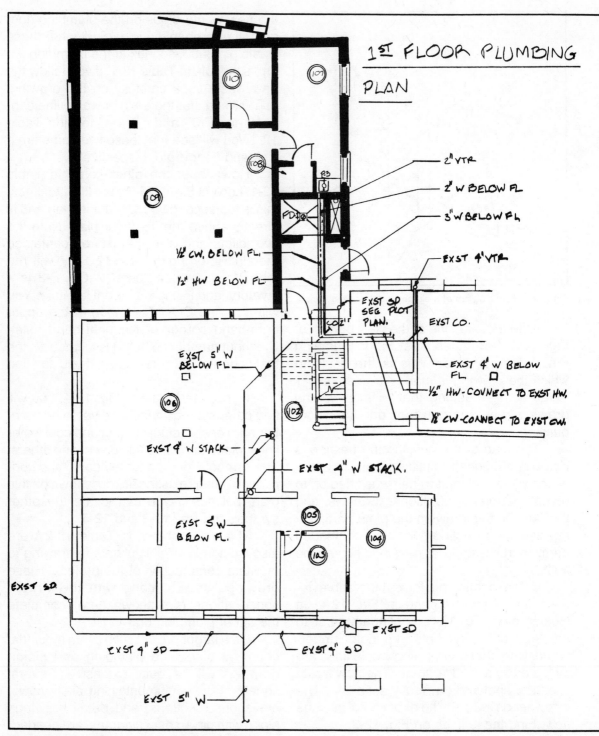

Fig. 12-3. Building floor plan showing plumbing layout.

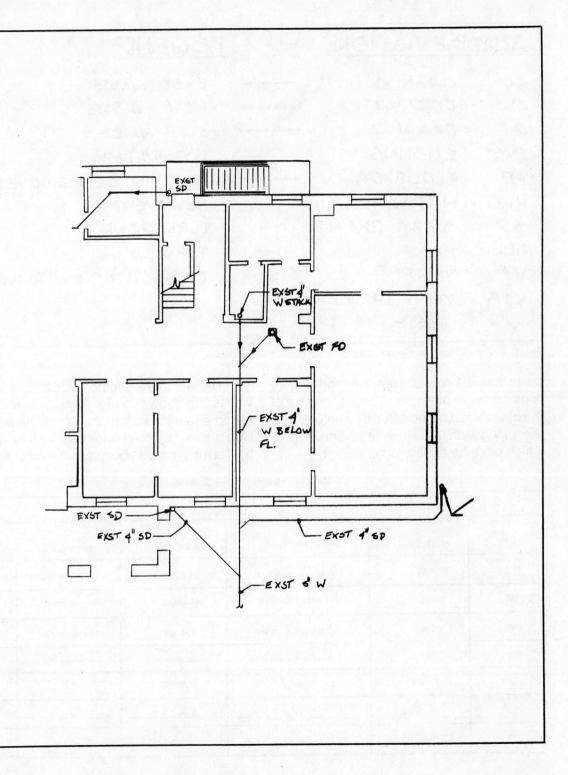

169

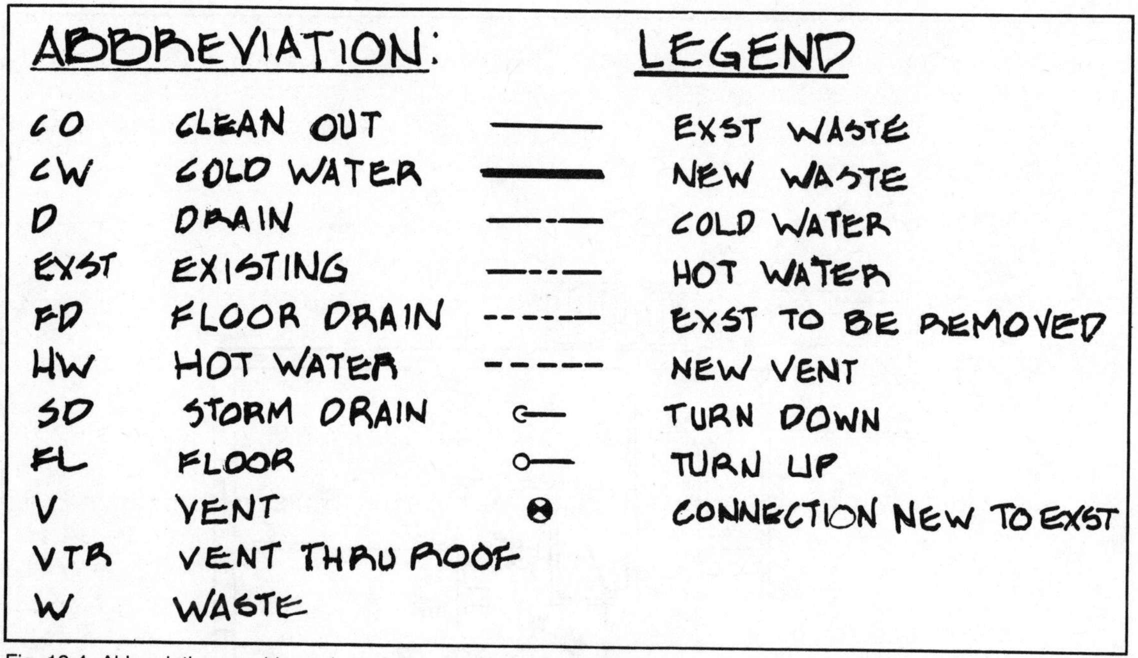

Fig. 12-4. Abbreviations and legend used on a set of plumbing drawings.

up to about 1 foot in diameter and double lines for larger pipes. In all cases the size of the pipe should be indicated by a note next to the pipes with an arrowhead showing to which pipe the dimension is meant.

The standard symbols for fittings and valves are shown in Fig. 12-6. You should become familiar with them. When the occasion arises, you may refer back to this symbol list for symbols you may have forgotten.

Table 12-1. Plumbing Fixture Schedule.

DESIG.	FIXTURE	MANUFACTURER	MODEL NO	MATERIAL	
P-1	W.C.	American standard	2108.058	Vitreous china	
P-2	Lav	American standard	0361.055	Vitreous-china	
P-3	Sink	American standard	6064.216	Stainless steel	
P-4	Lav.	American standard	9140.013	Vitreous china	
FD-1	Floor dr.	Josam	30802-5A-2	C.I.	
CO-1	Clean out	Josam	56024-2	C.I.	
CO-2	Clean out	Josam	58563-20	C.I.	

You will frequently find plumbing drawings drawn in an isometric view to make the picture clearer. To better understand this type of drawing, refer to Chapter 6.

HEATING, VENTILATING, AND AIR CONDITIONING DRAWINGS

If you read HVAC drawings, you will encounter many piping and sheet metal drawings. This is the most popular way of conveying heating and cooling from one location to another. Air ducts in your home carry hot or cold air to the various rooms in the house. Perhaps you have a steam or hot-water heating system. Pipes are used to carry the hot steam or water throughout the areas to be heated.

Heat Pump Operation

In recent years heat pumps have been used rather extensively in new homes and existing homes to replace less efficient energy sources. Solar heat is also being widely used, but few homes are heated exclusively by solar heat. Some other heat source is used as a backup. Heat pumps can be used as the sole heating (and cooling) system. Their use will reduce energy consumption as much as 30 to 60 percent without an enormous outlay of cash needed for the installation or conversion.

A *heat pump* is a system in which refrigeration equipment takes heat from a source and transfers it to a conditioned space (when heating is desired). The equipment removes heat from the space when cooling and dehumidification are desired.

The most common type is the *air-to-air heat pump*, where outside air is used as both a heat source and a heat-discharge medium. This type of heat pump can extract heat from the air at almost any temperature—even below zero—and move it inside the conditioned space.

Although heat pumps are more expensive to install than most conventional heating/cooling systems, they soon pay for themselves due to their ability to furnish more heat energy than they consume in electricity. Their useful heat output as compared to input is two or more to one—

PLUMBING FIXTURE SCHEDULE

SIZE	SUPPLY FITTINGS	TRAP	CARRIER	ACCESSORIES	CONNECTIONS		
					W	CW	HW
18"H	3405.016-⅜"	Integral	Floor	Seat 5320.114	4"	½"	——
20×18	2379.018-⅜"	1¼" P	Wall hung	——	1¼"	½"	½"
16¼×20¼	8310.070	1½"P	Counter	4331.013 Strainer	1½"	½"	½"
20×27	2238.129-⅜"	1½" P	Wall hung		1¼"	½"	½"
5"	——	Integral	——	——	2"	——	——
4"	——	——	——	——	4"	——	——
3"	——	——	——	——	3"	——	——

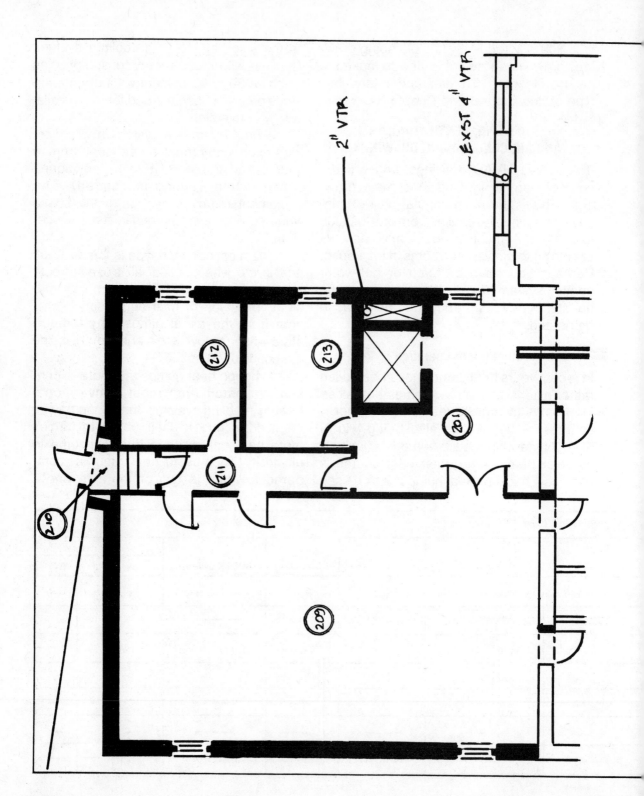

2" VTR

EXST 4" VTR

212

213

201

211

210

209

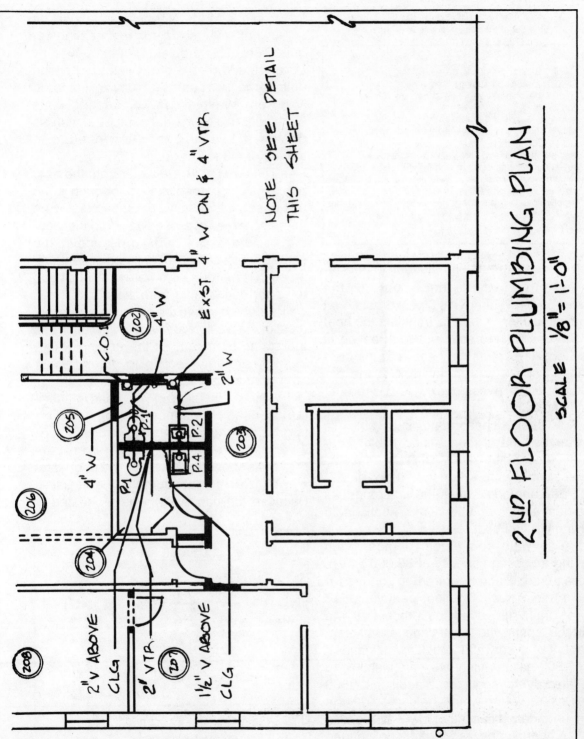

Fig. 12-5. Second floor plumbing plan for a building.

NOTES:
1. All exist conditions shall be field verified to ensure the proper installation of new equipment and services.
2. Insulate all exist and proposed HW piping inside bldg with ¾″ fiberglass with flame retardant vapor barrier jacket. See spec.
3. Insulate all proposed HW & CW piping exposed to outside conditions or earth with ¾″ rubatex foam insulation and two coats of waterproof paint. See spec.
4. Insulate all exposed fixture drains and their respective supply pipes in RM's 204 & 205 with ½″ rubatex foam.
5. See architectural sh. for storm drain details.

sometimes even as high as four to one.

Another type of heat pump is the *water-to-air* system. This type extracts heat from circulating water and transfers the heat to the conditioned space (when heating is desired). It extracts the heat from the conditioned space and transfers it to the circulating water (when cooling is desired).

The water-to-air system is especially useful when a cheap water supply (like a stream or surface spring) is available. A residence near Madison, Virginia, used a surface spring with a constant flow of 52-degree-Fahrenheit water (the year round) as the heat source and heatsink. A noticeable savings in fuel cost resulted.

Two water-to-air heat pumps, each having a cooling capacity of 36,000 Btu/h, were used. Each required a condenser water rate of about 8 gallons per minute (16 gpm total). As added insurance that the surface spring would provide a sufficient amount of water (after much investigation), two 1900-gallon reservoirs were installed to be filled by the spring. The additional 3800 gallons of water would help provide the amount needed in case of low water levels.

A heat pump sized for the cooling load

for buildings in the Northern states will be somewhat undersized for the heating requirement. This problem can be solved by installing add-on electric resistance heating units to the heat pumps. For the residence in Madison, Virginia, 9.6-kW heating units were added to each heat pump. Each unit is designed to energize when needed in stages of 4.8 kW each.

The used or waste water from the heat pumps is discharged into a floor drain in the basement. It then flows through a drainpipe and discharges at a headwall onto the sloping ground about 100 yards above the spring. This water is not polluted or contaminated. It is merely warm water that has been used for condensing purposes. The headwall is located above the spring in hopes that most of the waste water would flow back into the spring by means of the ground water-bearing stratum—cooling to ground water temperature as it flows (Fig. 12-7).

A similar installation was used in a residence near Harrisonburg, Virginia. This large home had a big indoor swimming pool. Two water-to-air heat pumps were connected as shown in Fig. 12-8. The pool water was used as a heat source and a heatsink. Because the pool water was preheated to

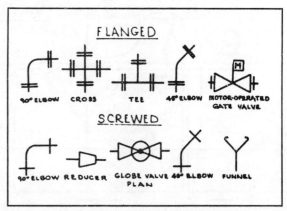

Fig. 12-6. Standard symbols for fittings and valves.

174

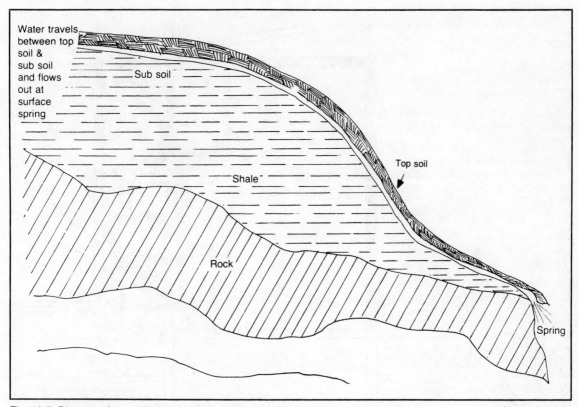

Fig. 12-7. Diagram of ground water-bearing stratum.

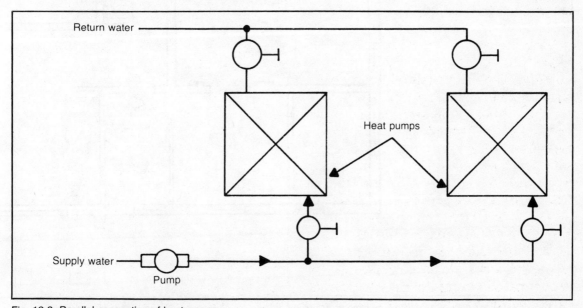

Fig. 12-8. Parallel connection of heat pumps.

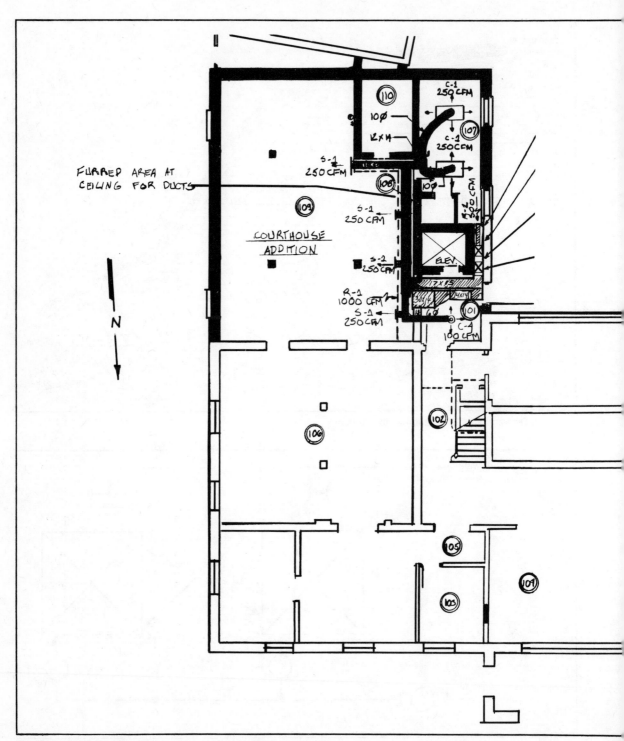

Fig. 12-9. HVAC floor plan of an existing building with a new addition.

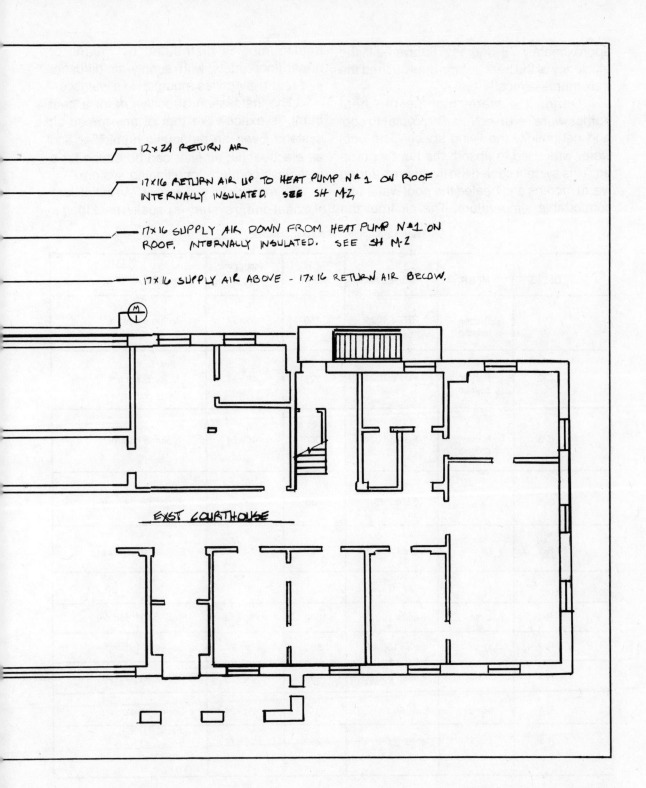

12 x 24 RETURN AIR

17 x 16 RETURN AIR UP TO HEAT PUMP Nº 1 ON ROOF
INTERNALLY INSULATED. SEE SH M-2,

17 x 16 SUPPLY AIR DOWN FROM HEAT PUMP Nº 1 ON
ROOF. INTERNALLY INSULATED. SEE SH M-2

17 x 16 SUPPLY AIR ABOVE - 17 x 16 RETURN AIR BELOW.

EXST COURTHOUSE

approximately 78 degrees Fahrenheit, the efficiency of the heat pumps approached the maximum—almost 4 to 1.

During the warm months the heat pumps were reversed (cooling cycle) to cool and dehumidify the living space. The pool water was used to absorb the heat extracted. This system cooled the living area during warm months and heated the pool water to a comfortable temperature. The air from the heat pumps is distributed by means of under-floor ducts with supply-air diffusers and return-air grilles mounted in a wainscot.

The installation of ductwork for a heat pump is exactly like that of any forced-air system. Even an automatic humidifier and an electrostatic air filter can be added for a first-class comfort-conditioning system.

As mentioned previously, the initial cost of a heat pump system is usually more than a

Table 12-3. Grille and Diffuser Schedule.

DESIG	MFGR	MODEL No	CFM	GRILLE SIZE	ACCESSORIES
C-1	Barber-colman	PBTV 1024	250	48×24	Volume control
C-2	Barber-colman	PBTV 1024	275	48×24	Volume control
C-3	Barber-colman	PBTV 1024	300	48×24	Volume control
C-4	Lima	Series 65	100	60	Series 68 damper
S-1	Lima	Series 100 VH	250	10×6	Series 101 damper
R-1	Lima	Series 100 HF	1000	30×14	Series 101 damper
R-2	Lima	Serie 100 VF	500	12×24	Series 101 damper
R-3	Lima	Series 90CC2	900	24×24	Series 90 damper
R-4	Lima	Series 90 CC 2 TB	1100	24×24	Series 90 damper
R-5	Lima	Series 90 RA VP	250	18×12	Series 90 RA VPT

Table 12-4. Abbreviations Used on a Set of HVAC Drawings.

CAP'Y	Capacity
C.I.	Cast iron
db	Dry bulb
ENT	Entering
Exst	Existing
EXT	External
FL	floor
MBH	1000 Btu/hour
SH	Sheet
wb	Wet bulb
ΔT	Change in temp
EQPT	Equipment
REQD	Required
T-STAT	Thermostat

comparable conventional heating and cooling system. You may also find that service is not the best in your locale. When heat pumps were first introduced on the market some years ago, all the "bugs" had not been removed. Many heating contractors became so discouraged with these units that they quit installing and servicing them altogether. Most of the problems have been solved, though, and many HVAC contractors are again installing and servicing heat pumps.

Working Drawings for Heat Pump Installation

Now that you know how a heat pump operates, let's see what various installations look like on paper—that is, working drawings to enable a complete installation.

The floor plan of an existing building and a new addition is shown in Fig.12-9. Note the drawing title—"1st Floor Plan"— and the scale that is ⅛ inch = 1 foot-0 inches. In this drawing the shaded partitions are the new addition, and those left plain are the partitions of the existing structure. Because the plan is of the first floor of a two-story building and the heat pumps are located on the roof, they are not shown. Only the ductwork can be seen.

Grilles and diffusers are designated by letters and numerals such as C-1, S-1, etc. The grille and diffuser schedule (Table 12-3) accompanied the set of drawings for this project. It lists the manufacturer, model number, cfm's, grille size, and accessories for all the grilles and diffusers shown on this set of HVAC drawings. A legend of abbreviations for this project is shown in Table 12-4.

The ductwork shown in Fig. 12-9 may be sufficient for the very experienced workman to install properly. It leaves plenty to be desired for those not experienced in heating installations and even for the experienced workman. Look at the roof plan in Fig. 12-10. This drawing is made as if someone was looking straight down at the top of the building from a taller building or else in an airplane. The viewer would see the tops of the two roof-mounted heat pumps. These are shown approximately to scale on this drawing. The construction details showing how the ductwork (shown on the floor plan) connects to these heat pumps are still not too clear.

A section of the building in question is shown in Fig. 12-11, and another is in Fig. 12-12. These details, along with the second floor plan in Fig. 12-13, help the workmen properly install the heat pump.

The heat pump schedule in Table 12-5 gives all the data necessary to know what heat pumps are required, along with any necessary installation data. Sometimes supplemental information may be included in the written specifications. These drawings, after adding the notes in Table 12-6, should enable anyone with a basic knowledge of sheet metal work and HVAC systems to properly install the system as designed by the HVAC engineer and shown on the working drawings.

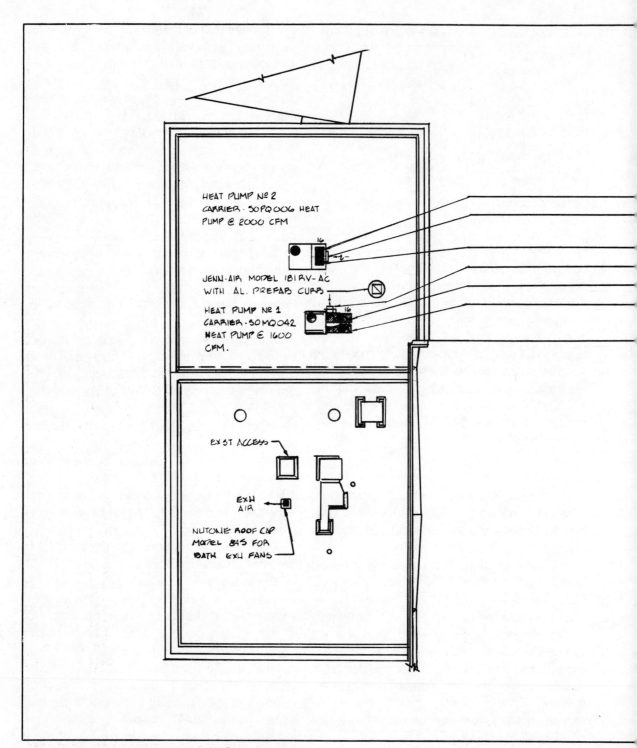

HEAT PUMP № 2
CARRIER · 50 PQ 006 HEAT
PUMP @ 2000 CFM

JENN·AIR MODEL 181 RV-AC
WITH AL. PREFAB CURBS

HEAT PUMP № 1
CARRIER · 50 MQ 042
HEAT PUMP @ 1600
CFM.

EX'ST ACCESS

EXH
AIR

NUTONE ROOF CAP
MODEL 845 FOR
BATH EXH FANS

Fig. 12-10. Roof plan used on a set of HVAC drawings.

16 x 18 RETURN THRU ROOF

OA INTAKE SET @ 250 CFM
WITH ½' x ½' BIRD SCREEN.

16 x 14 SUPPLY THRU ROOF

O.A. INTAKE SET @ 250 CFM, 10' x 6" WITH ½ x ½" BIRD SCREEN.

16 x 17 RETURN THRU ROOF

16 x 17 SUPPLY THRU ROOF

PART ROOF PLAN

SCALE: ⅛" = 1'-0"

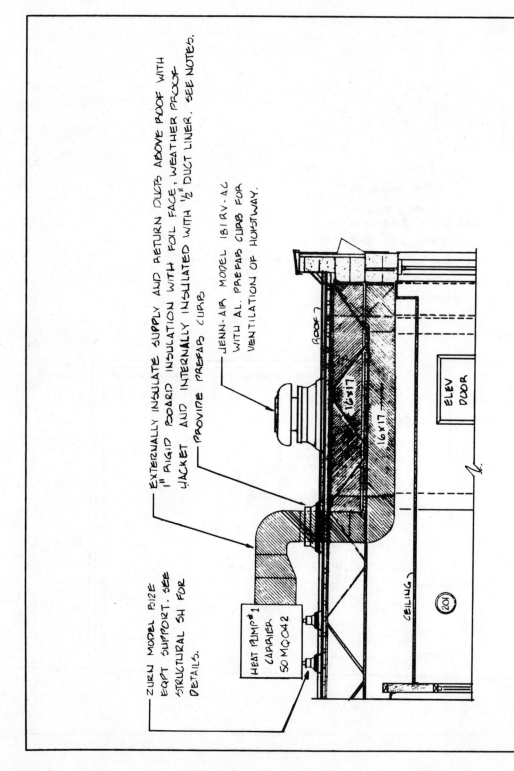

EXTERNALLY INSULATE SUPPLY AND RETURN DUCTS ABOVE ROOF WITH 1" RIGID BOARD INSULATION WITH FOIL FACE, WEATHER PROOF JACKET AND INTERNALLY INSULATED WITH ½" DUCT LINER. SEE NOTES.

JENN-AIR MODEL 1B1RV-AC WITH AL. PREFAB CURB FOR VENTILATION OF HOISTWAY.

PROVIDE PREFAB CURB

ZURN MODEL 1512E EQPT SUPPORT. SEE STRUCTURAL SH FOR DETAILS.

HEAT PUMP #1 CARRIER 50 MQ042

16×17

16×17

ROOF

ELEV DOOR

CEILING

201

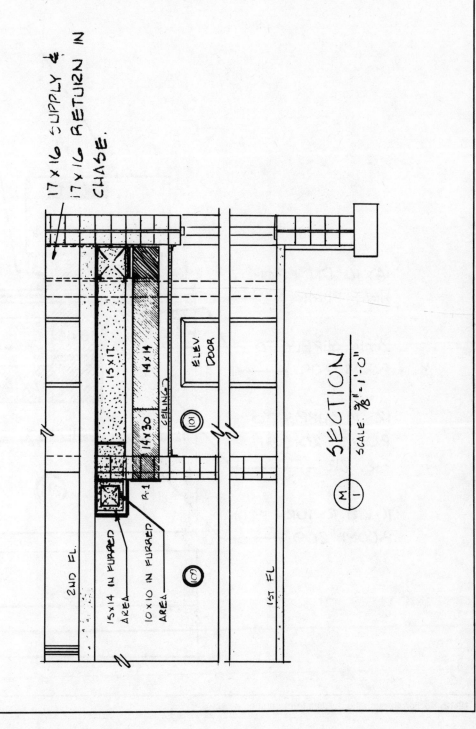

Fig. 12-11. Section through the building under discussion showing arrangement of HVAC equipment and ductwork.

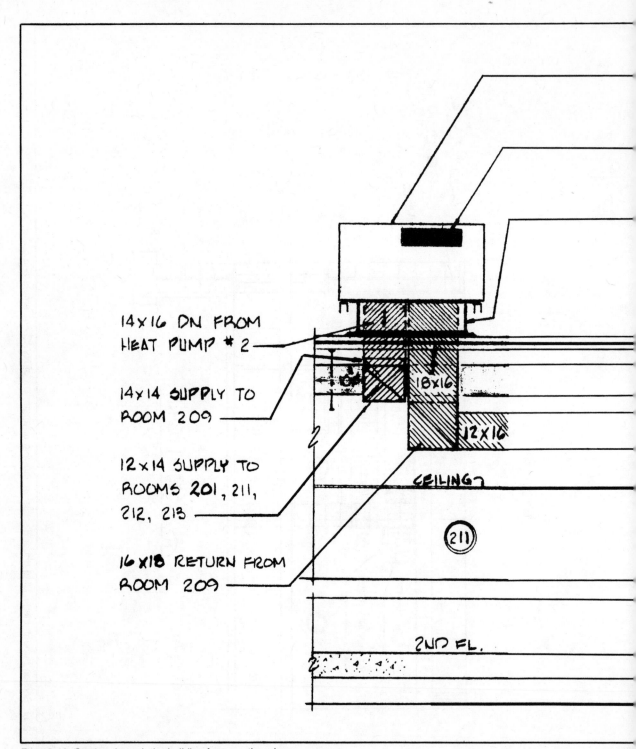

14x16 DN FROM
HEAT PUMP #2

14x14 SUPPLY TO
ROOM 209

18x16

12x16

12x14 SUPPLY TO
ROOMS 201, 211,
212, 213

CEILING

211

16x18 RETURN FROM
ROOM 209

2ND FL.

Fig. 12-12. Section through the building from another view.

184

HEAT PUMP № 2 · CARRIER · 50 PQ006
HEAT PUMP @ 2000 CFM.

O.A INTAKE SET @ 250 CFM. PROVIDE
½" x ½" BIRD SCREEN.

MANUFACTURER'S ROOF TOP CURB.

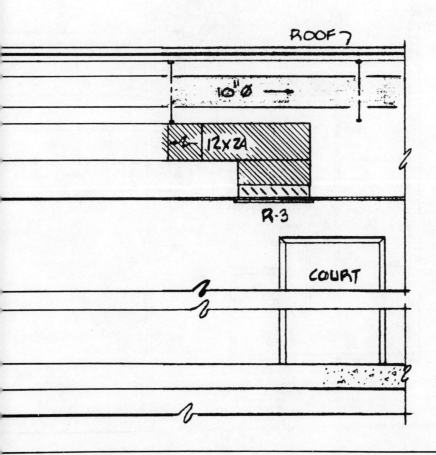

ROOF

10" ∅

12 x 24

R·3

COURT

185

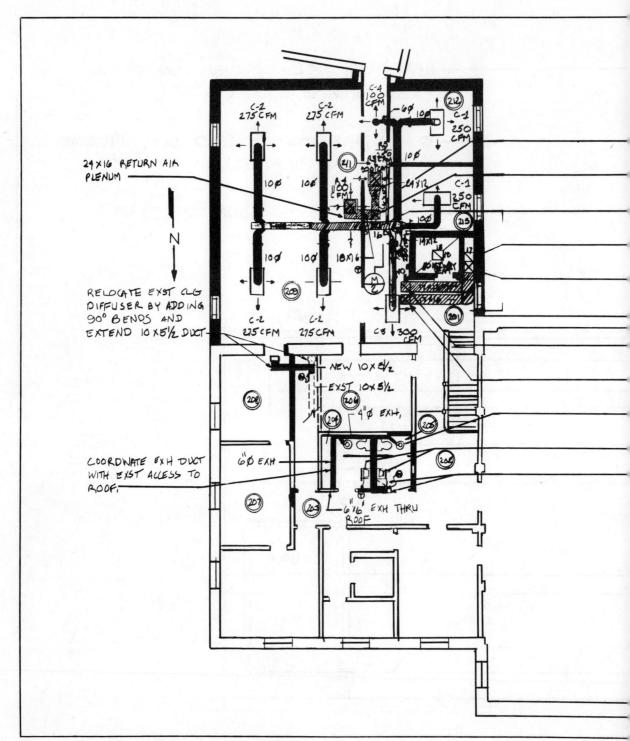

Fig. 12-13. Second floor plan in the set of HVAC drawings.

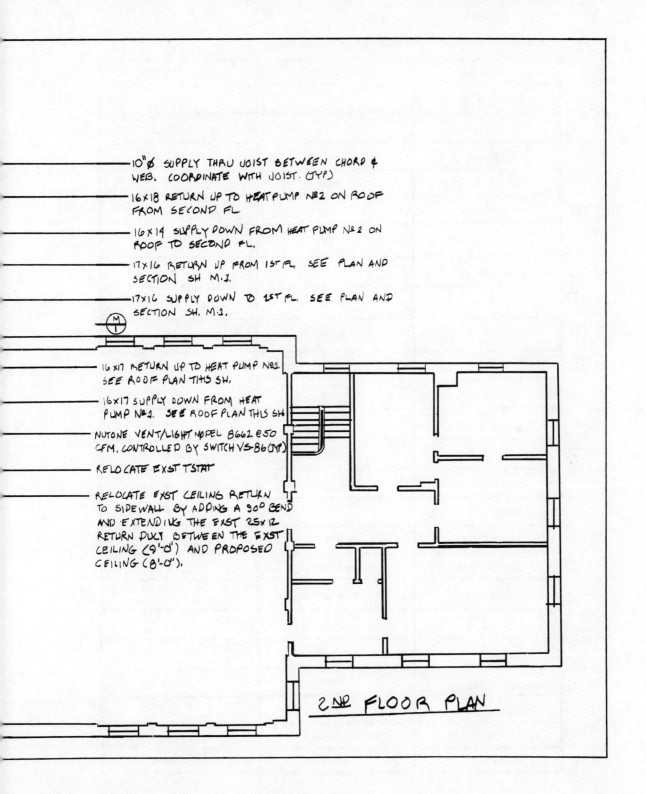

10"∅ SUPPLY THRU JOIST BETWEEN CHORD & WEB. COORDINATE WITH JOIST. (TYP.)

16X18 RETURN UP TO HEAT PUMP №2 ON ROOF FROM SECOND FL.

16X14 SUPPLY DOWN FROM HEAT PUMP №2 ON ROOF TO SECOND FL.

17X16 RETURN UP FROM 1ST FL. SEE PLAN AND SECTION SH M.2.

17X16 SUPPLY DOWN TO 1ST FL. SEE PLAN AND SECTION SH. M.1.

16X17 RETURN UP TO HEAT PUMP №1. SEE ROOF PLAN THIS SH.

16X17 SUPPLY DOWN FROM HEAT PUMP №1. SEE ROOF PLAN THIS SH.

NUTONE VENT/LIGHT MODEL 8662 @50 CFM. CONTROLLED BY SWITCH VS-86 (TYP)

RELOCATE EXST TSTAT

RELOCATE EXST CEILING RETURN TO SIDEWALL BY ADDING A 90° BEND AND EXTENDING THE EXST 25X12 RETURN DUCT BETWEEN THE EXST CEILING (9'-0") AND PROPOSED CEILING (8'-0").

2ND FLOOR PLAN

Table 12-5. Heat Pump Schedule.

DESIGNATION	HEAT PUMP NO 1	HEAT PUMP NO 2
Manufacturer	Carrier	Carrier
System type	Packaged	Packaged
Model No	50 MQ 042	50 PQ 006
Indoor fan capy. CFM	1600	2000
Total sp-in. H_2O		
Ext. sp-in. H_2O	.59	.50
Fan motor H.P.	½	¾
Volts/phase	200/3	200/3
Drive	Direct	Direct
Cooling coil	Heat pump	Heat pump
Ent. air db-°F	80° F	80°F
Ent. air wb-°F	67°F	67°F
Total cooling-MBH	42.0	59.0
Heating medium	Heat pump	Heat pump
Ent. air db- °F	10°F	10°F
Total capy-MBH	18.3	29.0
Elec resist heat	88EH0050MA00	Minimum
Volts/phase-KW	208/3 5KW	200/3 4.4KW
Total capy-MBH	17.1	15.0
Compressor-KW	5.7	7.9
Volts/phase	208/3	200/3
Temp air in condenser-°F	95°F	95°F
Filter	Integral	Integral
Outside air.	250 CFM	250 CFM

Table 12-6. Notes Used on a Set of HVAC Drawings.

NOTES

1. All exst conditions shall be field verified to insure the proper installation of eqpt.
2. All ducts in concealed spaces, except those with internal insulation shall be externally insulated with 1½″ duct wrap with flame safe vapor proof jacket. See spec.
3. Internal insulation shall be ½″ fiberglass duct liner.
4. Exposed ducts on roof shall be externally insulated with 1″ fiber. Glass rigid board with foil face vapor proof jacket in addition to internal insulation.
5. See architectural sheet 8 for reflective ceiling plan to coordinate installation of diffuser & grille

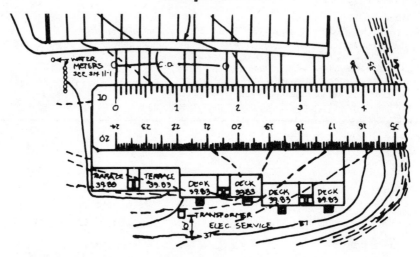

Chapter 13

Machine Shop Drawings

MACHINE SHOP DRAWINGS ARE BASED ON A "language" by which a designer expresses his thoughts to the mechanic or other workmen. This type of drawing may represent construction of a simple tool or a complicated machine.

A drawing for the machine trades ideally should be so informative and complete that the workmen can take it and, without further verbal or written instructions, produce the object as the designer intended it to be made. The information contained in a machine shop drawing should cover the form and size of the object, the kind of material from which it is to be made, the number of pieces desired, and the finish of its surfaces.

Most machine shop drawings will fall under one or more of the following categories: working drawings, assembly drawings, tool drawings, installation drawings, and manufacturing drawings. Most machine shop drawings involve construction of items such as jigs and fixtures.

JIGS AND FIXTURES

Jigs are used in performing identical operations easily and rapidly with uniformity of precision. Jigs are specifically adapted for such operations as drilling, reaming, counterboring, and tapping. A jig is used to hold the workpiece in position and to guide the cutting tool.

Fixtures are also locating and holding devices. Unlike jigs, they are clamped in a fixed position and are not free to move or guide the cutting tool. Most fixtures are used to perform operations requiring facing, boring, milling, grinding, welding, and the like.

The following rules are recommended by Brown and Sharpe Manufacturing Company for the designing and producing of jigs and fixtures.

●Before laying out the jig or fixture, decide on the locating points and outline a clamping arrangement.

●Make all clamping and binding devices as quick-acting as possible. These devices may be purchased from various sources and should be used whenever possible.

●See that two component parts of a machine can be located from corresponding points and surfaces.

●Make the jig foolproof. Arrange it so that the work cannot be inserted except in the correct way.

●Make some of the locating points adjustable for rough castings.

●Locate clamps so that they will be in the best position to resist the pressure of the cutting tool during the operation.

●If possible, make all clamps integral parts of the jig or fixture.

●Avoid complicated clamping arrangements that may wear or get out of order.

●As nearly as possible, place all clamps opposite from the bearing points of the work to avoid springing.

●To make the tools as light as possible, core out all unnecessary metal.

●Round all corners.

●Provide handles wherever they will make the handling of the jig more convenient.

●Provide feet (preferably four) opposite all surfaces containing guide bushings in drilling and boring jigs.

●Place all bushings inside of the geometrical figure formed by connecting the points of location of the feet.

●Provide sufficient clearance, particularly for rough castings.

●If possible, make all locating points visible to the operator when placing the work in position.

●Provide holes or escapes for the chips.

●Locate clamping lugs to prevent springing of fixtures on all tools that must be held to the table of the machine. Provide tongues that will fit slots in the machine tables in all milling and planing fixtures.

●Use ASA drill jig bushings—namely, head press fit bushings and slip renewable bushings on all new tool designs.

●Use liners where slip renewable bushings are used.

●Use headless liners for all general applications.

Shop drawings of simple devices are often confined to a single assembly that incorporates the essential details. For more complicated devices, both an assembly drawing and parts detail drawings are prepared.

TAP WRENCH

Most assembly drawings are prepared with sufficient views so that the assembly is clearly understood. Sectional views are often necessary to illustrate the construction. Figure 13-1 is a typical assembly drawing and details of a tap wrench. The very top view in the upper left-hand corner is the side view of the wrench the one immediately under it is a plan view. It is drawn as though the wrench is lying flat on a table and the viewer is looking straight down on it. Other details on the drawing include a section of the body, spring, and adjusting sleeve.

The sectional view of the body indicates that the wrench is made of steel, along with the adjusting sleeve. When you examine the plan view, note that the plunger fits in the end of the body. The spring fits between this

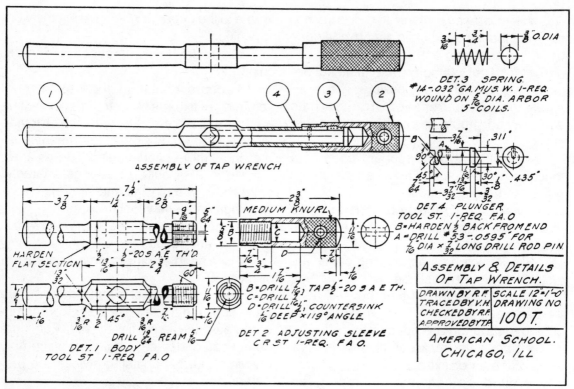

Fig. 13-1. Assembly drawing of a tap wrench.

plunger and the wrench body. Notes also indicate that the spring is to be made from number 14—.032-inch gauge music wire. Because the adjusting sleeve is threaded, it

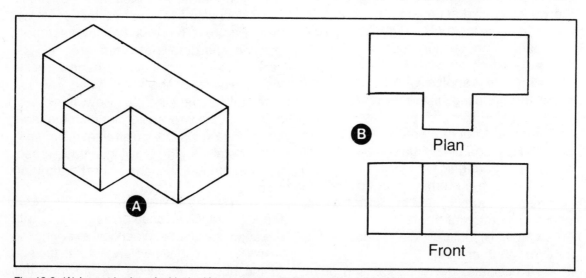

Fig. 13-2. (A) Isometric view of a block with a projection (B) Plan and front view of the object shown in A.

may be adjusted by turning it either clockwise or counterclockwise.

Figure 13-1 has few parts. Assembly of this wrench may be done without using reference numbers. If many parts were required in the assembly, it would be compulsory to use reference numbers on the drawing to facilitate reading it. Reference numbers are thus included in Fig. 13-1 to give you an idea of how such numbers are used and arranged. Reference numbers 1, 2, 3, and 4 on the drawing are referred to in the notes as body, adjusting sleeve, spring, and plunger, respectively.

Letters are also used in this drawing to convey information. On the plan view you will find that the dimensions are not specified for the tapped and drilled holes. The tapped hole is specified by B, and the drilled hole is indicated by C. Notes on drawings give necessary details.

The object in Fig. 13-2A is shown in an isometric view. The mechanical drawing in Fig. 13-2B requires two views to fully describe the object. Two solid lines are used to represent the projection on the block in the front view, and the actual shape of the projection is shown in the plan or top view.

V BLOCK

Figure 13-3A shows an isometric view of a V block. The vertical lines in the front view represent the edges of the V block's various parts. A top view is required to complete all the facts regarding these lines. The middle line in the front view represents the corner in the bottom of the V. Again, two views are required to relate all the facts concerning the object.

BRASS RING

Figure 13-4 shows a working drawing of a brass ring that is 3 inches in diameter and ⅛ inch thick. The thickness of the ring is given on the end view. The large hole in the ring is 2 inches in diameter. This dimension is found on the front view. There are four small holes drilled through the ring that are equally spaced. The size of these holes is indicated

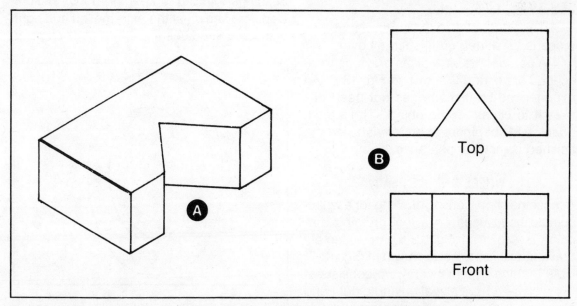

Fig. 13-3. (A) Isometric drawing of a V block. (B) Top and front view of the block shown in A.

193

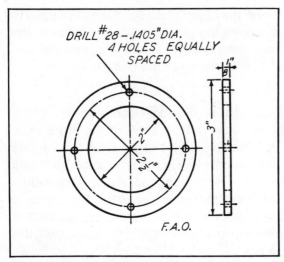

Fig. 13-4. Working drawing of a brass ring.

directly above the drawing, which is drill number 28—.1405 inch in diameter. These holes are located on a circle that is 2½ inches in diameter. If you want to know whether or not these holes are drilled all the way through the ring, look at the end view. The holes are drilled all the way through, because dotted lines are drawn from one side of the view to the other.

The symbol F.A.O. means that this piece is machined or finished all over.

A perspective view of the previously described brass ring is shown in Fig. 13-5. An experienced craftsman does not need this sketch to construct the object. The sketch enables the beginner to better visualize the finished object's appearance.

WHEEL SPINDLE HEAD

A machine shop working drawing for a wheel spindle head used on a tool grinding machine is shown in Fig. 13-6. An assembly drawing of this project is shown in Fig. 13-7. Note that the assembly drawing consists of two views—a front elevation and a right end elevation. In the former view the upper part is

shown in section, the central part is a full view, and the lower part is partly indicated. In the right end elevation the central part is fully shown, and the lower part is omitted. Practices vary in making assembly drawings. Some assembly drawings have all the piece parts indicated by numbers, thereby aiding the mechanic to assemble the pieces. Piece parts or detail numbers are not specified in this case.

At the top of the front elevation or center part of the machine, some of the pieces are enclosed. The piece that covers them is the wheel spindle head. A spindle rotates within this wheel spindle head, which has bearing housings on each end with bushings inside the housings. This spindle holds the emery wheels. A pulley is located inside this wheel spindle head and between the bearings which, as will be seen from the section, is one made of cast iron and held to the spindle by a headless setscrew. Two pieces are shown in the section to the right and left of the pulley. These pieces are babbitted bushings that aid in keeping down friction. Caps are shown at the extreme left and right of the babbitted bushings. The caps come in

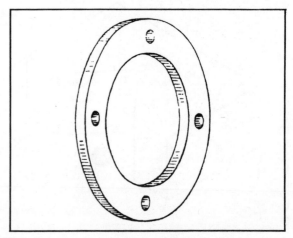

Fig. 13-5. Perspective view of the brass ring shown in Fig. 13-4.

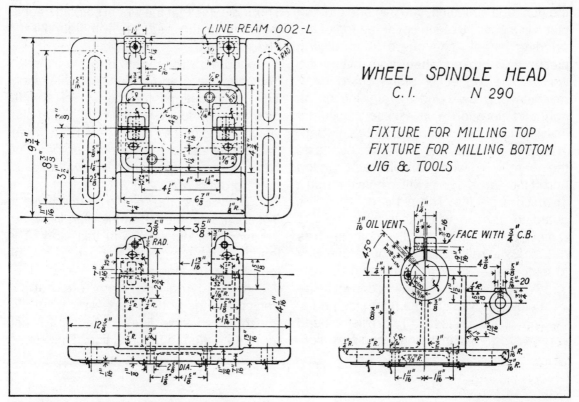

Fig. 13-6. Shop drawing for a wheel spindle head.

contact with the right and left bearings. These caps are fastened to the bushings with headless setscrews.

You can gain a better understanding of the spindle head from a careful study of the right end elevation. The part of the head that is attached to the pedestal is grooved out somewhat at the top. This is indicated by invisible or dotted lines.

To the right and left of the machine's centerline are lines tapering up from the bottom of the head to the bottom of the bearing. These lines indicate the sides of the ribs. Refer to the sectional part of the front elevation. The perpendicular lines directly below the bearings and setting back slightly from the bearings' inner ends indicate the length of the ribs.

The upper part of the bearing housings are not crosshatched in the front sectional elevation. Refer to the right end elevation at the upper part of the housings. The front sectional elevation was obtained by taking a cutting plane on the centerline in the right end elevation. Here two invisible perpendicular lines may be seen, one to the right and one to the left, extending from the top of the bearing down to the bushings. You will also find to the right and to the left of the top part of the housings a screw passing through, which indicates that the housings are split at the top.

In the front elevation at the extreme right end of the spindle, a certain piece directly above the sectional elevation is very vague in this particular view. If you refer to the right end elevation, you can see that the piece is a guard used to cover the top of the grinding wheel. The guard keeps flying pieces of a broken wheel from hitting the worker. The connection of this guard to the machine may be found by referring to the right end elevation. You will see visible and invisible lines connecting to a circular part. The lines indicate that a bracket connects the guard to the wheel spindle head. On the end of the arm is a boss with a hole through it in which a pin fits. The pin is used for the guard to turn on when it is necessary to remove the wheel from the spindle. It is necessary to raise the guard out of the way to remove the wheel from the spindle.

A flat fillister head screw is shown in the front elevation view on the right side and at the top of the pedestal. Two fillister head screws are shown in the right end elevation.

This indicates that the wheel spindle head is fastened to the pedestal with four fillister head screws.

Figure 13-6 consists of a plan view, a front elevation, and a right end elevation. The specifications for the wheel spindle head may be found on the right of the plan view; C.I. indicates that cast iron is the material, and N 290 is the pattern number. The notes refer to special tools available for mass production.

The width of the spindle head is given at 9-¾ inches on the left of the plan view, but the length is not shown in this view. The length dimension, 12-⅝ inches, can be found by referring to the front elevation.

In studying the front elevation—particularly the bottom part—you will find a horizontal invisible line located 7/16 inch from the bottom horizontal line. This shows

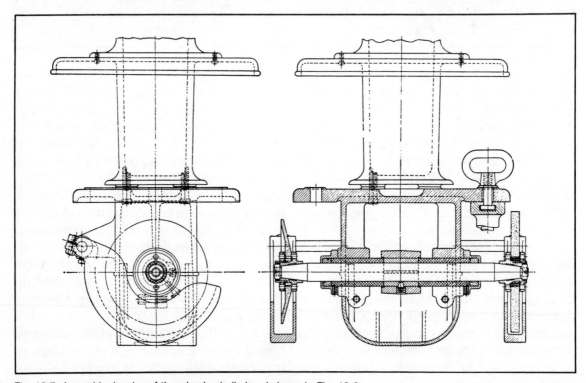

Fig. 13-7. Assembly drawing of the wheel spindle head shown in Fig. 13-6.

that the base has an impression in it, but the length and width of the impression are not given. The person building the object will make the top part of the base 2-5/8 inches on each end and then allow for the length of the impression. The impression starts a short distance from the back and front side and terminates at the perpendicular line. The extreme width of the boxlike design on the casting is shown. The correct location cannot be obtained from this particular view. A 1/4-inch thickness is specified in the plan view.

You will find from Fig. 13-7 that the housings are supported by ribs directly below them. To determine the distances from the outer edge of the vertical part of the casting to the edge of the rib, refer to Fig. 13-6 between the right housing and the base. The distance is 1-5/16 inches. You will also find near the front elevation directly to the side of the right housing that the distance from the center of the housing to the top of the base is 4-1/16 inches. On some blueprints this dimension would be specified on the right side elevation; on others it would be on the front elevation.

The housings are located on the base. Refer to the front elevation to the right of the perpendicular centerline, and note that the dimension is 1-13/16 inches. This is the distance from the centerline to the inside face of the bearing housing. This dimension is used for locating the other housing. For the length of the housing, add 1-13/16 inches to 1-5/8 inches, the result is 3-7/16 inches. Subtract 3-7/16 inches from 3-5/8 inches. The sum is 3/16 inch. Add 3/16 inch to 1-5/8 inches. The total is 1-13/16 inches. To the right of the left housing is the dimension 2-3/4 inches, which is the outside diameter of

the housing. The dimension of the diameter of the hole in the housing is 1-5/8 inches.

After looking at the front elevation of the working drawing, it is difficult to determine whether or not parts of the housings are enclosed in a boxlike casting. By studying the right-hand elevation, however, you will see that the bearings are partly encased in a rectangular boxlike casting. Refer to the plan view. The corners of this box are rounded, and the radius of these corners must be found before they can be properly constructed. Below the right housing in the front right-hand corner of the box, the radius is shown as 3/8 inch, which is the radius to use for the inner corner. The outside radius, as seen from the rear right-hand corner of the box, is 5/8 inch.

On the right end elevation directly above the housing, you will find lugs that are to be machined through the center with a 1/8-inch saw. Their design or shape is not indicated. Refer to the front elevation and study the various parts of the bearing housing. The housing to the left shows that the ear or lug has a radius of 1/2 inch. Not knowing just how high the top part of this lug is to come, again read the blueprint. You will not find the distance from the center of the housing to the center of the hole in the ear specified on the front elevation. Referring to the right end elevation, you will find that this dimension is 9/16 inches. This gives the correct position to set dividers when scribing the arc that will give the rounding for the ear.

Find out how the grinding wheel guard is supported. This guard swivels on a pin inserted in a hole in a bracket that is at the back of the boxlike part of the casting (Fig. 13-7). Find the location of the bracket relative to the top and center of the box. As it is the center of the bracket in which the hole is machined, look for the location of the horizontal centerline. On the right end elevation (Fig. 13-6) above the bracket and the boxlike part of the casting is the dimension 4-⅜ inches. This is the specification for the horizontal distance. Between the bearing and the box's right is the dimension 1½ inches—wall the distance from the top of the box to the center of the bracket in which the hole will be machined.

Chapter 14

Structural Drawings

STRUCTURAL DRAWINGS ARE ALWAYS PRESENT IN plans for houses and other buildings. These drawings normally show footings, foundation, structural framing, and other structural details. The drawings are often prepared by structural engineers based on proper allowances for all vertical loads and lateral stresses of the building in question.

Footings are concrete "feet" placed in the ground and sometimes reinforced with steel bars on which the foundation and the subsequent building load are placed. Soil conditions, together with the weight of the building, determine the size, design, and number of footings. Building weight is measured in terms of *dead* and *live loads*. Dead load is the stationary weight of the building itself and the permanently fixed furnishings and equipment. Live load is made up of movable equipment in the building and people who use it.

Foundation walls serve as a base on which a building is built. The walls carry the load of the building to the footings and earth below. These walls naturally must be strong enough to resist the side pressure of the earth. Foundation walls may be built of concrete block, stone, reinforced concrete, or similar materials. Another function of the walls is to keep moisture out of the underground parts of the building. Usually a waterproofing compound is applied to the exterior surface of foundation walls. Drain tile, combined with gravel, is also used to divert the water away from the building's interior.

Most of the larger buildings constructed in recent years have frameworks of steel beams. Even the smaller commercial buildings that are predominantly constructed of masonry block or brick utilize steel-truss construction for roofs. Cranes, bridges, antenna towers, machinery structures, and other objects are composed of steel members fastened together by bolts, rivets, or welds to form a strong structure. Structural drawings usually show the dimensions of the steel pieces and the method in which they are assembled.

Although structural drawings encompass other materials than steel, most structural drawings deal mainly with steel. To become competent in reading structural drawings, you should first become familiar with steel types. Steel is made from iron with the addition of carbon, silicon, nickel, manganese, molybdenum, and chromium to make it stronger, more workable, and more resistant to rust. The steel types have been tested and coded, so the proper type for a given job is easily determined.

The required strength of a structure depends on location, size, type, and the weight and size of the material the structure has to support. Calculating the shapes and sizes of the structural members is normally done by a certified structural engineer. He will specify the type or types of steel required, along with dimensions and how each type is fastened to another member. A code number usually is used to indicate the various steel types—for example, ASTM A7. The letters ASTM stand for American Society for Testing and Materials. All codes can be found in code books published by the ASTM.

STEEL SHAPES

There are numerous rolled steel shapes currently in use. Most rolling mills issue catalogs containing drawings and data pertaining to each type of product that they produce. Four types have been used more often than others. These are the *plate, angle, I beam,* and *channel.* The Z bar, T bar, rod,

flat, oval, bulb angle, bulb beam, and numerous other rolled shapes are also used, but not as extensively as the first four types mentioned.

The plate consists of a rectangular prism of rolled steel having width, thickness, and length. The views of a plate are shown in Fig. 14-1. The elevation is the principal view. The end view is a view of the end, assuming the end is to be rotated through 90 degrees into the plane of the paper. Section AA shows the conventional method of indicating a section. The arrowheads indicate the place where the assumed section is taken. The arrowheads indicate the direction in which the section is viewed.

The size of a plate is specified by dimensions or a schedule giving the width in inches or millimeters, the thickness in inches or fraction of an inch, and the length in feet and inches—unless the metric system is used.

The angle is a unit of rolled steel whose section resembles an L. The two parts are called "legs" and are of uniform thickness. There are at least two sets of angles—those having legs of equal widths and those having legs of unequal widths. Structural views of an angle are shown in Fig. 14-2. The elevation is the principal view in this drawing, and

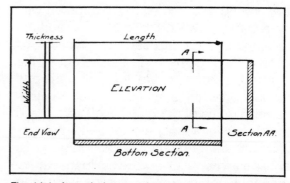

Fig. 14-1. A steel plate consists of a rectangular prism of rolled steel.

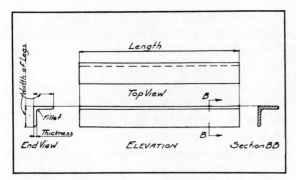

Fig. 14-2. The angle is a piece of rolled steel whose section resembles an L.

the others are grouped around it. It is not customary to show the fillets and curved edges of angles and other shapes in structural steel drawings. This would take too much time. The size of the angle is specified by giving the width and thickness of legs in inches and the length in feet and inches—or equivalent metric dimensions.

The I beam is a rolled steel shape whose cross section resembles the letter I. The upper and lower portions are called the *flanges*, and the central portion is the *web*. The inside of the flange has a slope of 2 inches in 12 for all standard I beams. These beams may be obtained in various depths, widths of the flange, and thicknesses of the web. Figure 14-3 shows several views of an I

beam based on the principal elevation. Note that the flanges in the elevation are represented by two dotted or "invisible" lines rather than three. These lines are drawn to scale and are somewhat thicker than the outside edge of the flange. The size of the I beam is specified by giving the depth in inches, the weight per foot in pounds, and the length in feet and inches.

The channel is that steel shape whose cross section is shaped like the I beam with one side of each flange removed. The parts are named as in the I beam, and the slope of the inside of the flange is the same. The structural views are shown in Fig. 14-4. Channels may be obtained in various depths, widths of flange, and thicknesses of web as with the I beam. The size of the channel is specified by giving the depth in inches, the weight per foot in pounds, and the length in feet and inches.

Figure 14-5 gives the top and end view along with the elevation of a T bar. The top portion is called the flange, and the lower portion is the *stem*. The size of a T bar is specified by giving the width of the flange in inches, the depth of the stem in inches, the weight per foot in pounds, and the length in feet and inches.

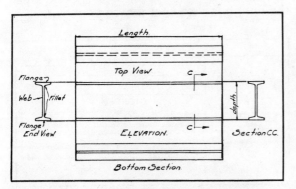

Fig. 14-3. The I beam is a rolled steel shape whose cross section resembles the letter I.

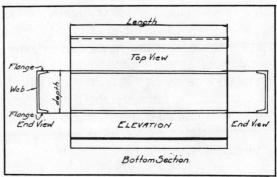

Fig. 14-4. The channel has its cross section shaped like the I beam with one side of its flange removed.

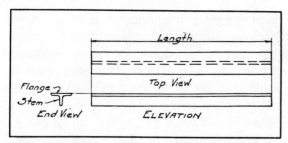

Fig. 14-5. The T bar's cross section resembles the letter T.

A Z bar is shown in Fig. 14-6. It is specified by giving the depth first in inches, the width of the flange next in inches, the thickness in inches, and the length in feet and inches. The thickness of the Z can be given because of its uniformity throughout.

The previous sketches show the various shapes of steel in several different views. On actual drawings you will seldom see them drawn in all possible views; nor is it necessary. Only views that are essential to give a clear representation of the structure to be fabricated will be seen on working drawings. There is no need to continue once the required views have been drawn. Don't expect steel shapes on actual drawings to be exactly as shown here.

FASTENING METHODS

There are several ways to fasten structural steel shapes together, but the most conventional ways include welding, rivets, bolts, soldering, or a combination of these methods. Regardless of the method used, the operation of properly proportioning these connections develops into a vital one. It is of little or no avail to have members of a structure well proportioned and of ample strength if the connections are too weak to carry the required loads.

Welding

This process involves first melting two pieces of steel to be joined at a desired spot and then fusing them with a third material. This method provides the strongest type of fastening, but it is also very costly. Five basic joints are used in welding—*butt, fillet, lap, edge,* and *corner welds*. The butt weld and fillet weld are the two most commonly used (Fig. 14-7).

Butt welds are made by placing two plates side by side, leaving 1/16 inch to 1/8 inch between them to get deep penetration. The plates should first be tacked at both ends. Otherwise, the heat will cause the plates to move apart (Fig. 14-8). The plates are then welded together from left to right. The electrode is pointed down in the crack—between the two plates—and is also slightly tilted in the direction of travel. The molten metal should be distributed evenly on both edges and in between the plates.

When fillet welds are made, you must hold the electrode in a 45-degree angle between the two sides, or the metal will not distribute itself evenly. Sometimes a holder is used to ensure that the electrode will be kept at a 45-degree angle. Multiple pass welds are frequently used. The first bead is melted in the corner with a fairly high current. The electrode is held at an angle to deposit the filler beads as shown in Fig. 14-9, putting the final bead against the vertical plate.

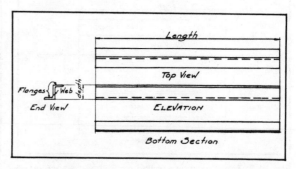

Fig. 14-6. As the name implies, a section of the Z bar is shaped like the letter Z.

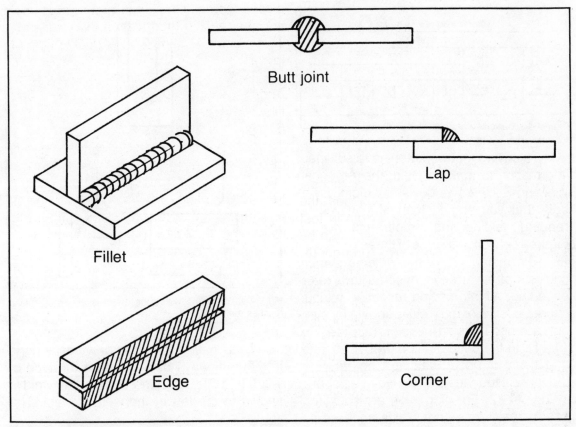

Butt joint

Lap

Fillet

Edge

Corner

Fig. 14-7. Various welding joints.

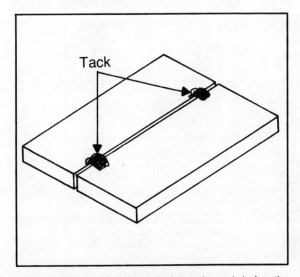

Tack

Fig. 14-8. Pieces should be tacked near the ends before the main weld begins to prevent the pieces from separating.

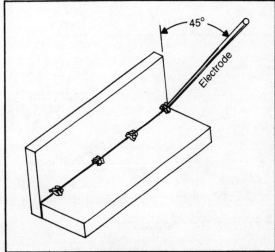

45°

Electrode

Fig. 14-9. The electrode should be held at a 45-degree angle to deposit the filler beads during a corner weld.

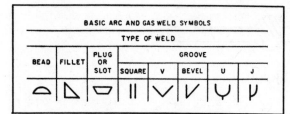

BASIC ARC AND GAS WELD SYMBOLS							
			TYPE OF WELD				
BEAD	FILLET	PLUG OR SLOT	GROOVE				
			SQUARE	V	BEVEL	U	J
⌒	△	▽	‖	∨	⦦	⋃	⋃

Fig. 14-10. Basic and arc gas weld symbols.

To ensure that workmen have sufficient information to make a weld, the American Welding Society has devised welding symbols that you should learn if you will be reading welding and structural drawings. Figure 14-10 shows some of these symbols.

The main base of the welding symbol consists of an arrow connected to a reference line. All information reflecting various characteristics of the weld is shown by symbols arranged around the reference line (Fig. 14-11). In the case of fillet, groove, flange, and flash or upset welding, the arrow is drawn so it touches the weld joint as shown in Fig. 14-12. For plug, slot, arc-spot, arc-seam, resistance-spot, resistance-seam, and projection welding, the arrow should touch the centerline of the weld joint (Fig. 14-13).

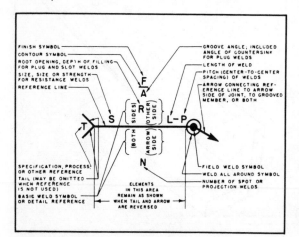

Fig. 14-11. On working drawings all information reflecting various characteristics of the weld is shown by symbols.

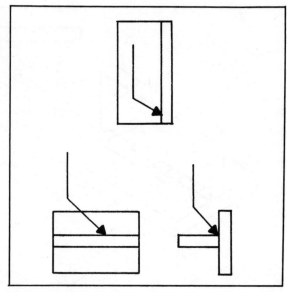

Fig. 14-12. In some welds the arrow is drawn so that it touches the weld joint.

Designating the sizes of welds is done in different ways depending on the type of weld. In a fillet weld the width of the weld is shown to the left of the weld symbol. The

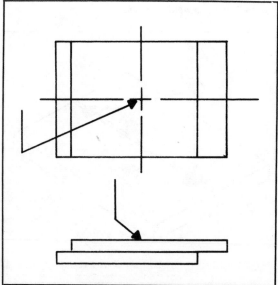

Fig. 14-13. For other welds, the arrow should touch the centerline of the weld joint.

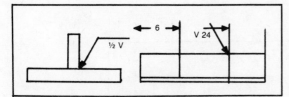

Fig. 14-14. In a fillet weld the width of the weld is shown to the left of the weld symbol. The length of the weld is shown to the right of the weld symbol.

length of the weld is shown to the right of the weld symbol (Fig. 14-14). The size of a plug weld is shown to the left of the weld symbol, and the center-to-center spacing is shown to the right of the weld symbol. Figure 14-15 shows how welding symbols are used in an actual construction drawing.

Rivets and Riveted Connections

In structural steel work the diameters of rivets in common use are ½ inch, ⅝ inch, ¾ inch, ⅞ inch, and 1 inch. The rivets are made

¾ inch in diameter in almost all building work, although sometimes light trusses are made with ⅝-inch rivets. For extremely large trusses and girders in large skyscrapers, ⅞-inch and sometimes 1-inch rivets are used. Tables giving the maximum size of rivets that should be used through the flanges of channels, I beams, etc., can be found in any standard handbook of structural steel. The holes in the metal through which the rivets pass are usually punched 1/16 inch.

The rivets commonly used in structural work consist of a cylindrical shank and a buttonhead. The rivets are heated and inserted in place. A second buttonhead is formed on the other end by a cup under pressure. There are times where the buttonhead will not fit. The head interferes with some other part of the structure. The buttonhead may be flattened to ⅜ inch or ¼

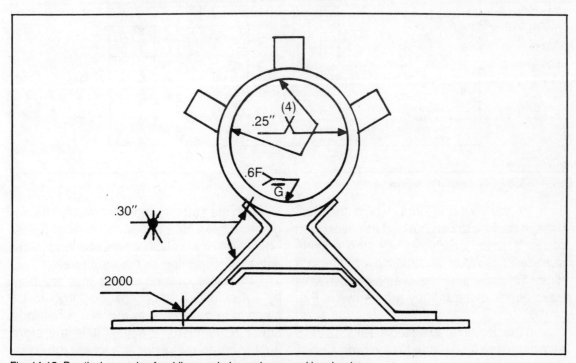

Fig. 14-15. Practical example of welding symbols used on a working drawing.

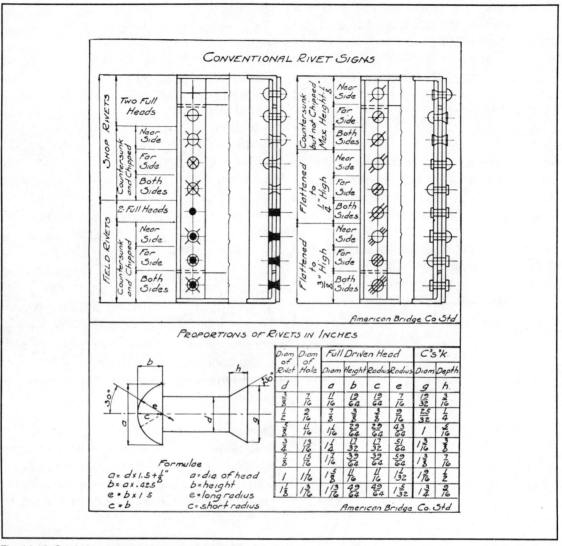

CONVENTIONAL RIVET SIGNS

		Shop Rivets	
	Two Full Heads		
Countersunk and Chipped	Near Side		
	Far Side		
	Both Sides		

Countersunk but not Chipped Max Height 1/8"

Flattened to 1/4" High

Flattened to 3/8" High

Near Side / Far Side / Both Sides / Near Side / Far Side / Both Sides / Near Side / Far Side / Both Sides

Field Rivets — 2-Full Heads — Countersunk and Chipped — Near Side / Far Side / Both Sides

American Bridge Co Std

PROPORTIONS OF RIVETS IN INCHES

Diam of Rivet	Diam of Hole	Full Driven Head				C's'k.	
d		Diam	Height	Radius	Radius	Diam	Depth
		a	b	c	e	g	h
$\frac{3}{8}$	$\frac{7}{16}$	$\frac{11}{16}$	$\frac{19}{64}$	$\frac{19}{64}$	$\frac{7}{16}$	$\frac{12}{32}$	$\frac{3}{16}$
$\frac{1}{2}$	$\frac{9}{16}$	$\frac{7}{8}$	$\frac{3}{8}$	$\frac{3}{8}$	$\frac{9}{16}$	$\frac{25}{32}$	$\frac{1}{4}$
$\frac{5}{8}$	$\frac{11}{16}$	$1\frac{1}{16}$	$\frac{29}{64}$	$\frac{29}{64}$	$\frac{43}{64}$	1	$\frac{5}{16}$
$\frac{3}{4}$	$\frac{13}{16}$	$1\frac{1}{4}$	$\frac{17}{32}$	$\frac{17}{32}$	$\frac{51}{64}$	$1\frac{3}{16}$	$\frac{3}{8}$
$\frac{7}{8}$	$\frac{15}{16}$	$1\frac{7}{16}$	$\frac{39}{64}$	$\frac{39}{64}$	$\frac{59}{64}$	$1\frac{3}{8}$	$\frac{7}{16}$
1	$1\frac{1}{16}$	$1\frac{5}{8}$	$\frac{11}{16}$	$\frac{11}{16}$	$1\frac{1}{32}$	$1\frac{9}{16}$	$\frac{1}{2}$
$1\frac{1}{8}$	$1\frac{3}{16}$	$1\frac{13}{16}$	$\frac{49}{64}$	$\frac{49}{64}$	$1\frac{5}{32}$	$1\frac{3}{4}$	$\frac{9}{16}$

American Bridge Co. Std.

Formulae

$a = d \times 1.5 + \frac{1}{8}"$ a = dia of head
$b = a \times .425$ b = height
$e = b \times 1$ e = long radius
$c = b$ c = short radius

Fig. 14-16. Symbols used to designate various types of rivets.

inch, or the metal through which the rivet passes may be drilled out and the rivet head made flush or nearly so with the outside surface. The latter is called a *countersink* rivet. To save time on drawings, symbols have been devised. They are shown in Fig. 14-16.

In the design of a riveted steel structure, allowance must be made for the reduction of area caused by punching or drilling holes through the steel member for rivets. The engineer allows for a certain number of holes. This number should not be exceeded in the actual construction of the structure.

Practically all structures that are built by a fabricating shop are too large to be shipped in one piece to the site where the structure is to be erected. It is therefore necessary to subdivide the whole into convenient members that can be readily handled and shipped. These members have a shipping or erection mark, usually a capital letter, painted upon them. They are sent to the site accompanied by a diagram showing where each member is to be placed. The members are then assembled into a finished structure. The various parts are bolted together and finally riveted up. Each member may consist of a single beam, channel, angle, and the like or, as is more often the case, it may be built up of several component parts. These parts are assembled and riveted together before the members leave the shop. This latter type of riveting is called *shop riveting*, while that done in the field or at the site is known as *field riveting*.

Most rivets are driven by a compressed-air machine. The rivet is heated to a red color and inserted in the hole. The machine is centered on the heated end while pressure is applied to form a neat buttonhead. Field riveting is usually done by the use of a pneumatic hand-hammer. The compressed air is supplied by a portable compressor and then through a hose to the hammer. On some small projects the rivets may be driven by hand with the use of a dolly, snap, and sledge, but this method is all but obsolete.

Bolting

High-strength steel bolts have replaced rivets in many instances for certain steel structures. The technique is very similar to riveting except no heat is used to form a buttonhead. Instead, a nut is screwed onto the threads of the bolt to secure the structural members together. Complete specifications for high strength steel bolts may be found in the pamphlet "Assembly of Structural Joints Using High Strength Steel Bolts," distributed by the American Institute of Steel Construction.

Soldering

Soldering differs from welding in that welding requires the melting of two pieces of steel at the jointing place, whereas soldering uses a third piece of material to be melted, which is allowed to flow between and around the two pieces of steel. When a very low melting solder is used for fastening, it is called soft soldering. Soldering techniques are used only on small low-stress structures such as a chassis for an electronic device and similar items.

COLUMN SCHEDULE

Construction crews on large buildings use structural drawings along with the architectural drawings to construct the building. These drawings will involve plans, elevations, sections, and large details dealing with entire floors or the whole building.

Plans for steel framing members of buildings are drawn for each floor and for any other levels that might have special steelwork. Most will also have a *column schedule*, which shows the location and numbering of all vertical columns. These columns should also be indicated in all plan views to show the exact location of each in relation to the building structure.

REINFORCED CONCRETE DRAWINGS

Two principal sets of drawings are used in reinforced concrete structures. One type is used to show the general arrangement of the structure and the size and reinforcement of the members. The other type shows the size, shape, and location of all steelwork in the structure. Drawings for reinforced concrete structures follow the same general layout as other types of structural drawings in that they include a plan, elevation, and

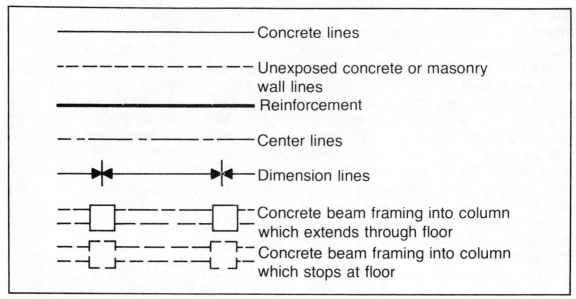

Fig. 14-17. Symbols used on reinforced concrete drawings.

section and detail views along with schedules of footings, columns, beams, and slabs.

Symbols and abbreviations are used on reinforced concrete drawings. A list of symbols is in Fig. 14-17. The various parts of a building are indicated by marks that designate the floor, type of members, and identification of a specific member. For example, 3B4 would mean third floor beam number four. The letters used to show the types of members are:

B Beams	L Lintels
C Columns	S Slabs
F Footings	T Ties
G Girders	U Stirrups
J Joists	W Walls

Chapter 15

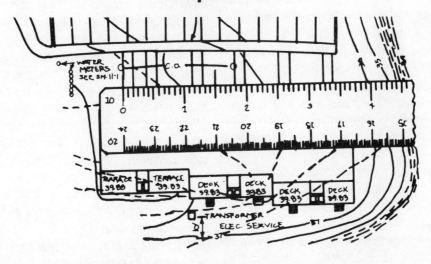

Architectural Drawings

IN ALL LARGE CONSTRUCTION PROJECTS AND IN most of the smaller ones, an architect is commissioned to prepare complete working drawings and specifications for the project. These drawings usually include:

●A plot plan indicating the location of the building on the property.

●Floor plans showing the walls and partitions for each floor or level.

●Elevations of all exterior faces of the building.

●Several vertical cross sections to indicate clearly the various floor levels and details of the footings, foundation, walls, floors, ceilings, and roof construction.

●Large-scale detail drawings showing such construction details as may be required.

For projects of any consequence, the architect usually hires consulting engineers to prepare structural, electrical, plumbing, heating, ventilating, and air conditioning drawings.

PLOT PLAN

This type of plan of the building site is as if the site is viewed from an airplane and shows the property boundaries, the existing contour lines, the new contour lines, the new contour lines (after grading), the location of the building on the property, new and existing roadways, all utility lines, and other pertinent details. Descriptive notes may also be found on the plot plan listing names of adjacent property owners, the land surveyor, and the date of the survey. A legend or symbol list is also included so that anyone familiar with site plans can readily read the information.

FLOOR PLANS

The plan view of any object is a drawing

showing the outline and all details as seen when looking directly down on the object. It shows only two dimensions—length and width. The floor plan of a building is drawn as if a slice was taken through the building—about window height—and then the top portion was removed to reveal the bottom part where the slice was taken.

Let's say that we first wanted a plan view of a home's basement. The part of the house above the middle of the basement windows is imagined to be cut away. By looking down on the uncovered portion, every detail, partition, and the like can be seen. Likewise, imagine the part above the middle of the first floor windows cut away and a drawing made looking straight down at the remaining part. This would be called the first floor plan.

ELEVATIONS

A plan view may represent a flat surface, a curved surface, or a slanting one, but for clarification it is usually necessary to refer to elevations and sections of the house. The *elevation* is an outline of an object that shows heights and may show the length or width of a particular side. The accompanying illustrations show three views of elevation drawings for a house. Note that these elevation drawings show the heights of windows, doors, porches, the pitch of roofs, etc.—all of which cannot be shown conveniently on floor plans. Notice also that the end view shows the views of the house that we would notice if we stood directly in front of the house and looked toward it. From this view we are able to understand whether the roof is of the gambrel type or other type, and it is also possible to determine the pitch of the roof from the small triangle on the roof.

SECTIONS

A *section* or sectional view of an object is a view facing a point where a part of an object is supposed to be cut away, allowing the viewer to see the object's inside. The point on the plan or elevation showing where the imaginary cut has been made is indicated by the section line, which is usually a very heavy double dot and dash line. The section line shows the location of the section on the plan or elevation. It is, therefore, necessary to know which of the cutaway parts is represented in the sectional drawing when an object is represented as if it was cut in two. Arrow points are thus placed at the ends of the sectional lines.

In architectural drawings it is often necessary to show more than one section on the same drawing. The different section lines must be distinguished by letters, numbers, or other designations placed at the ends of the lines as shown in the accompanying illustrations, in which the sections are lettered A-A and B-B. These section letters are generally heavy and large so as to stand out on the drawings. To further avoid confusion, the same letter is usually placed at each end of the section line. The section is named according to these letters—that is, Section A-A, Section B-B, and so forth.

A longitudinal section is taken lengthwise while a cross section is usually taken straight across the width of an object. Sometimes, however, a section is not taken along one straight line. It is often taken along a zigzag line to show important parts of the object.

A section view, as applied to architectural drawings, is a drawing showing the building, or portion of a building, as though cut through, as if by a saw, on some imaginary line. This line may be either vertical (straight up and down) or horizontal. Wall sections are nearly always made vertically so that the cut edge is exposed from top to bottom. In some ways the wall section is one of the most important of all the drawings to a practical builder, because it answers the questions on how to build. The floor plans of a house show how each floor is arranged, but the wall sections tell how each part is constructed and indicate the material to be used.

To better understand how architectural drawings are used, let's look at some drawings used on an actual building. Figure 15-1 shows an existing building and a new addition. A legend directly below the floor plan enables the persons reading the drawings to distinguish between the different lines used. Note that rooms are numbered by the numerals being enclosed in a double circle. These room numbers serve many purposes. Look at the finish schedule (Table 15-1). The room numbers, or in this case space numbers, are shown in the very left-hand column. The next column to the immediate right is a "designation" column that lists the names of the various spaces. Then come the floor, base, wainscot, etc., columns.

Spaces 204 and 205 on the floor plan are rest rooms. Due to the relatively small scale at which these are drawn, the architect deemed it necessary to show a larger scale drawing of these areas to help the workmen better understand what takes place. See Fig. 15-2. Note the roof access shown by dotted lines on this drawing. The method of constructing this access is very vague, so a detail like the one shown in Fig. 15-3 is in order.

A floor plan of the stairs is shown in Fig. 15-4. Note the section marks on this plan—a circle enclosing letters and numerals encased with an arrowhead. Three sections are taken: B-8, C-8, and D-8. Drawings of these sections are shown in Figs. 15-5, 15-6, and 15-7, respectively.

While observing the sections discussed in the previous paragraph, you probably noticed other designations such as "F-8" enclosed in a circle. This indicates the stair tread detail shown in Fig. 15-8. Other details to be used with the plan and sections of the stairs are shown in Figs. 15-9 through 15-11.

INFORMATION ON ARCHITECTURAL DRAWINGS

Let's discuss the information that makes up a complete set of architectural drawings. As mentioned previously, the size and number of drawing sheets used in a set of working drawings will depend on the size and complexity of the project. Regardless of the size or type of building sheets used in a set of working drawings will depend on the size and complexity of the project. Regardless of the size or type of building, nearly all sets of drawings will generally contain the following items.

The first phase of any building program consists of gathering certain information from the building owners, the site on which the building is to be constructed, and from the city or town in which the building will be built. All regulations and ordinances must be followed. The architect will usually have a preliminary conference with the owners to determine space requirements and then analyze various factors of the building program, such as type of structure, location and orientation of the building, availability of utilities, topography, traffic pattern of the sur-

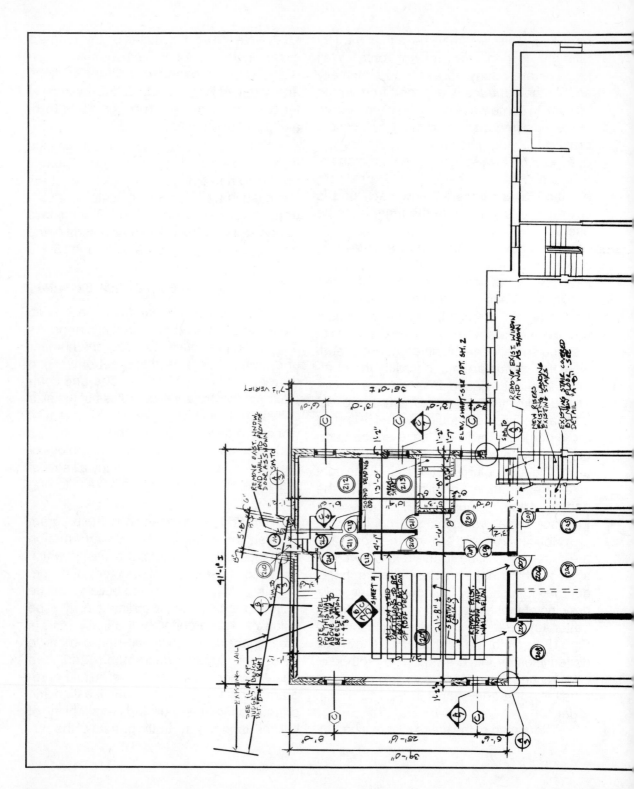

212

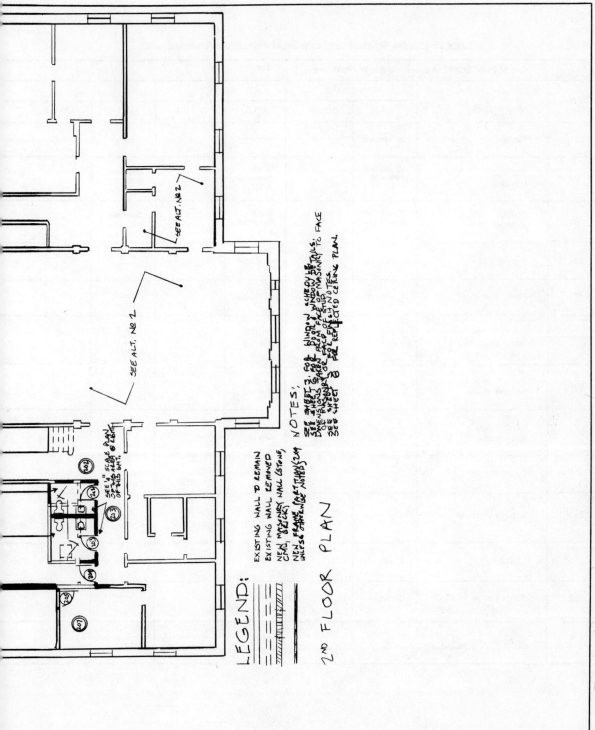

Fig. 15-1. Floor plan of an existing building showing the new addition.

213

Table 15-1. Finish Schedule Used on a Set of Architectural Drawings.

SPACE NO.	DESIGNATION	FLOOR		BASE		WAINSCOT	
		MAT.	COLOR	MAT.	COLOR	MAT.	COLOR
201	Lobby	V. asb. tile		Wood		——	
202	Stair hall (existing)	V. asb. tile See note # 4		Wood See note # 5		See note # 7	
203	Corridor (existing)	V. asb. tile See note # 4		Wood See note # 6		See note # 8	
204	Women's toilet	V. asb. tile		4" rubber		——	
205	Men's toilet	V. asb. tile		4" rubber		——	
206	J. & D.R. clerk (existing space)	V. asb tile & exist.		Exist. & New wd.		——	
207	G.D.CT. clerk (existing space)	Exist.		Exist. & new wd.		——	
208	G.D.CT. clerk (existing space)	Exist.		Exist. & new wd.		——	
209	Courtroom	Carpet		Wood		Painted ½" gyp. bd.	
210	Security vestibule	V. asb. tile		Wood		——	
211	Corridor	V. asb. tile		Wood		——	
212	Judge's office	Carpet		Wood		——	
213	Probation officer	V. asb. tile		Wood			

FINISH SCHEDULE See Sheet 3 for Finish Notes

WALLS		CEILING			
MAT.	COLOR	MAT.	COLOR	HGT.	REMARKS
Painted ½" gyp. bd.		Ac. tile		8'-0"	See note #3-note rubber treads stair addition
See note # 9		Exist. plaster		——	
See note # 9		Exist. ac. tile		9'-0"	See ¼" scale plan below. Cut exist. ac. tile CLG. suspension system, where wall is moved & furnish new wall angle. Furnish new ac. tile if necessary.
Painted ½" gyp. bd.		Ac. tile		8'-0"	See note #10. See ¼" scale plan below
Painted ½" gyp. bd.		Ac. tile		8'-0"	See note #10. See ¼" scale plan below
Exist. & painted ½" gyp. bd.		Exist. plaster		——	See note #11-repair exist. CLG. where partition is moved.
Exist & painted ½" gyp. bd.		Exist. plaster		——	See Note # 12
Exist. & painted ½" gyp. bd.		Exist. plaster		——	See note # 12
Painted ½" gyp. bd.		Ac. tile		8'-0"	See note #3
Painted ½" gyp. bd.		Painted gyp. bd.		8'-0"	Ceiling to be installed on 2×4 framing @16"O.C. enclose lintel @14" masonry wall ¾" wd. frame. Note rubber treads@ stairs.
Painted ½" gyp. bd.		Painted gyp. bd.		8'-0"	See note #13-ceiling to be installed on 2×4 framing at 16" O.C.
Painted ½" gyp. bd.		Ac. tile		8'-0"	See note #3 & 13
Painted ½" gyp. bd.		Ac. tile		8'-0"	See note #3 & 13

215

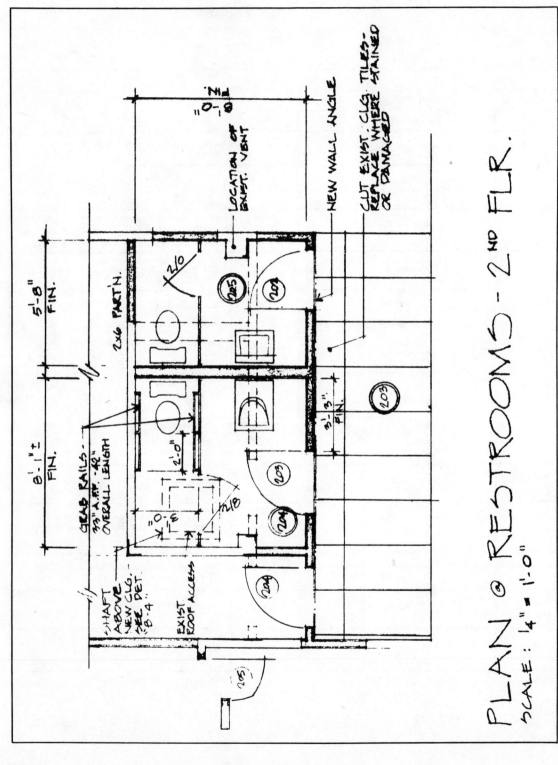

PLAN @ RESTROOMS - 2ᴺᴰ FLR.

SCALE : 1/4" = 1'-0"

Fig. 15-2. A larger scale drawing of two rest rooms.

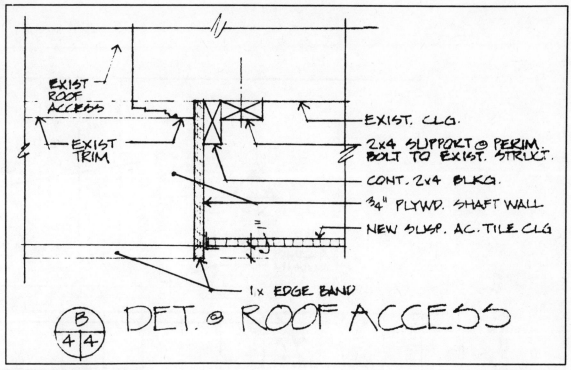

EXIST ROOF ACCESS

EXIST TRIM

EXIST. CLG.

2x4 SUPPORT @ PERIM. BOLT TO EXIST. STRUCT.

CONT. 2x4 BLKG.

¾" PLYWD. SHAFT WALL

NEW SUSP. AC. TILE CLG

1x EDGE BAND

DET. @ ROOF ACCESS

B 4/4

Fig. 15-3. Roof access detail.

rounding street, and the budget allotted for the particular project.

PRELIMINARY DRAWINGS

When the required information is obtained, the architect normally sketches a set of preliminary drawings showing a proposed floor plan layout and also elevations of the building. The architect may even furnish a roughly drawn perspective of the proposed building to give the owners an idea of what the finished structure will look like. Another conference is held to discuss the preliminary sketch, and usually many changes are made at this time. When the preliminary sketches and the accompanying notes seem to satisfy the owners, the architect may then begin work on the actual construction documents consisting of working drawings and written specifications.

CONSTRUCTION DRAWINGS

From the preliminary sketches and notes taken during the preliminary conferences, the architect and his draftsmen begin working on the construction drawings. These drawings must provide sufficient information so that along with the written specifications, no significant decision is left to the direction of the contractor. The set of construction documents will therefore include all essential building details such as plot plan, foundation plan, floor plans, elevations, sections, and as many clarification details and enlarged drawings as are necessary. Finished drawings are also prepared that detail the structural, electrical, heating, plumbing, and ventilating systems. For large buildings, these latter categories are usually prepared by consulting engineering firms.

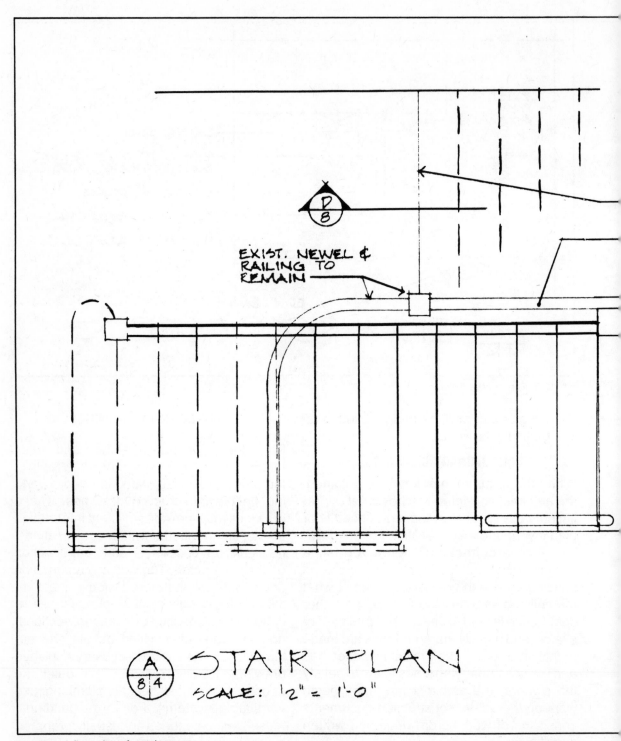

EXIST. NEWEL & RAILING TO REMAIN

STAIR PLAN
SCALE: 1/2" = 1'-0"

Fig. 15-4. A floor plan of a stairway.

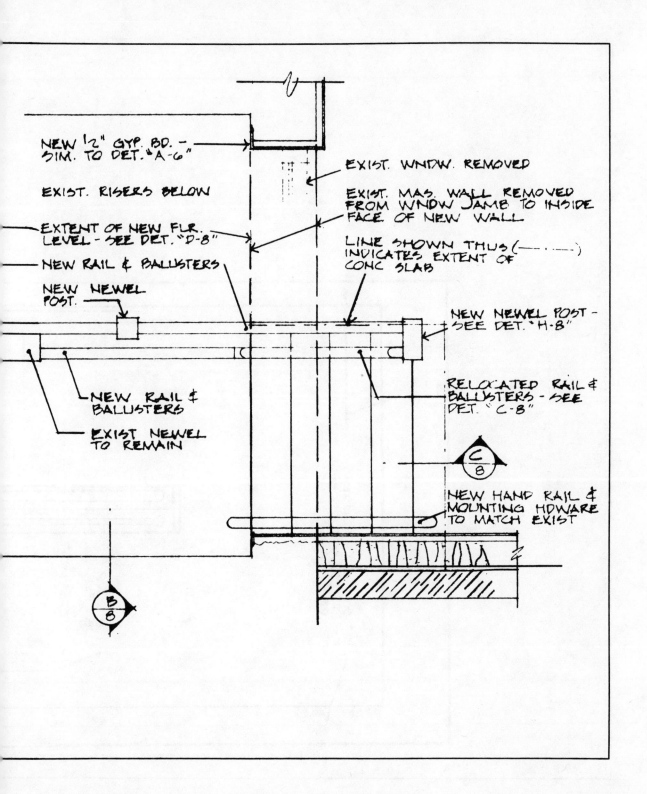

NEW $\frac{1}{2}$" GYP. BD. –
SIM. TO DET. "A-6"

EXIST. RISERS BELOW

EXTENT OF NEW FLR.
LEVEL - SEE DET. "D-8"

NEW RAIL & BALUSTERS

NEW NEWEL
POST.

NEW RAIL &
BALUSTERS

EXIST NEWEL
TO REMAIN

EXIST. WNDW. REMOVED

EXIST. MAS. WALL REMOVED
FROM WNDW JAMB TO INSIDE
FACE OF NEW WALL

LINE SHOWN THUS (—. —.)
INDICATES EXTENT OF
CONC SLAB

NEW NEWEL POST -
SEE DET. "H-8"

RELOCATED RAIL &
BALUSTERS - SEE
DET. "C-8"

NEW HAND RAIL &
MOUNTING HDWARE
TO MATCH EXIST

C / 8

B / 8

219

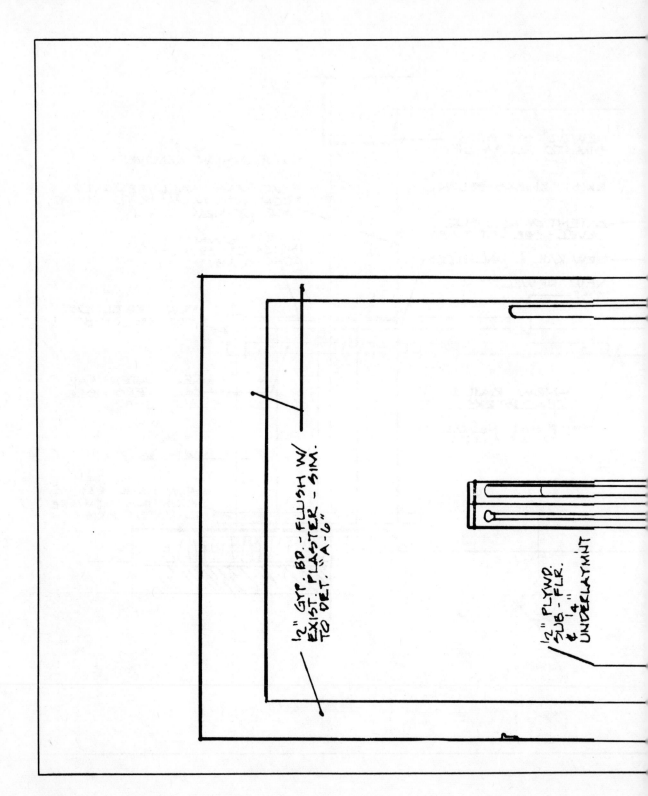

½" GYP. BD. - FLUSH W/
EXIST. PLASTER - SIM.
TO DET. "A-6"

½" PLYWD.
SUB - FLR.
& ¼"
UNDERLAYMNT

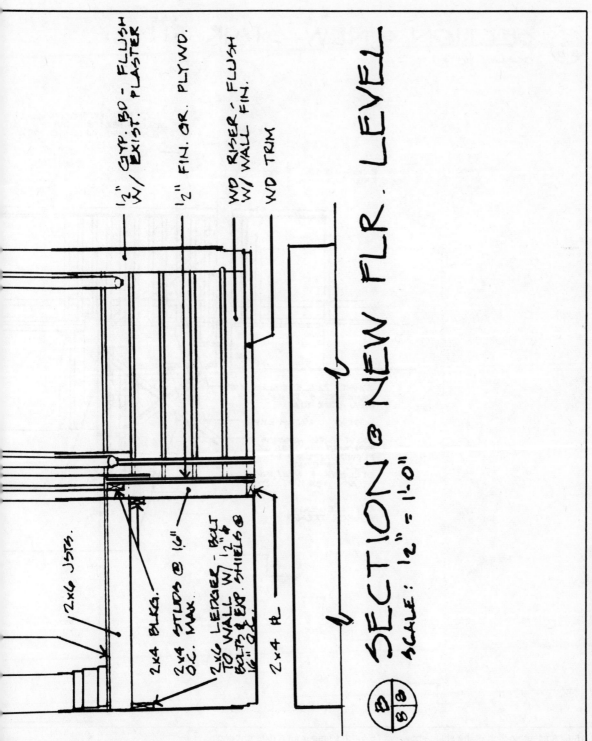

½" GYP. BD. - FLUSH
W/ EXIST. PLASTER

½" FIN. GR. PLYWD.

WD RISER - FLUSH
W/ WALL FIN.

WD TRIM

2×6 JSTS.

2×4 BLKG.

2×4 STUDS @ 16"
O.C. MAX.

2×6 LEDGER - BOLT
TO WALL W/ ½"∅
BOLTS & EXP. SHIELDS @
16" O.C.

2×4 FL.

B SECTION @ NEW FLR. LEVEL
8 SCALE: ½" = 1'-0"

Fig. 15-5. Section of the stairs shown in Fig. 15-4.

221

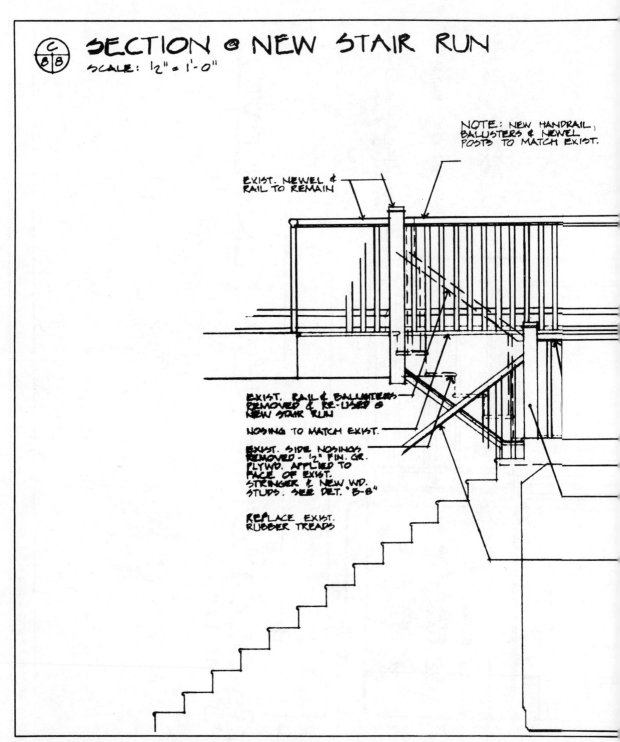

SECTION @ NEW STAIR RUN
SCALE: 1/2" = 1'-0"

C / 8

NOTE: NEW HANDRAIL, BALUSTERS & NEWEL POSTS TO MATCH EXIST.

EXIST. NEWEL & RAIL TO REMAIN

EXIST. RAIL & BALUSTERS REMOVED & RE-USED @ NEW STAIR RUN

NOSING TO MATCH EXIST.

EXIST. SIDE NOSINGS REMOVED - 1/2" FIN. GR. PLYWD. APPLIED TO FACE OF EXIST. STRINGER & NEW WD. STUDS. SEE DET. "B-8"

REPLACE EXIST. RUBBER TREADS

Fig. 15-6. Another section of the stairs shown in 15-4—this time from a different angle.

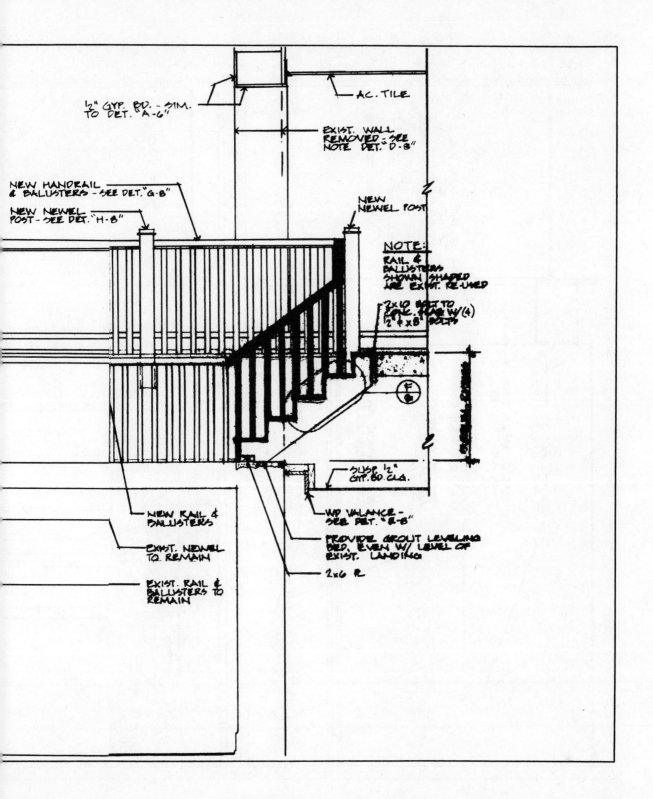

½" GYP. BD. - SIM. TO DET. "A-6"

AC. TILE

EXIST. WALL REMOVED - SEE NOTE DET. "D-8"

NEW HANDRAIL & BALUSTERS - SEE DET. "G-8"

NEW NEWEL POST - SEE DET. "H-8"

NEW NEWEL POST

NOTE: RAIL & BALUSTERS SHOWN SHADED ARE EXIST. RE-USED

2x10 BOLT TO CONC. SLAB W/ (4) 2 + x8" BOLTS

SUSP. ½" GYP. BD. CLG.

NEW RAIL & BALUSTERS

EXIST. NEWEL TO REMAIN

EXIST. RAIL & BALUSTERS TO REMAIN

WD VALANCE - SEE DET. "E-8"

PROVIDE GROUT LEVELING BED. EVEN W/ LEVEL OF EXIST. LANDING

2x6 R

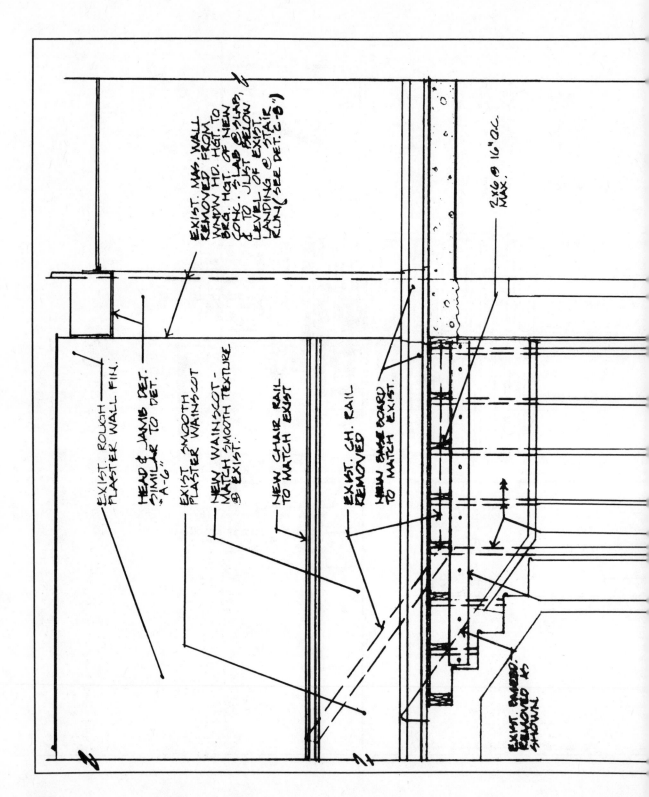

EXIST. MAS. WALL REMOVED FROM WNDW HD. HGT. TO BRG. HGT. OF NEW CONC. SLAB @ LNDG, & TO JUST BELOW LEVEL OF EXIST. LANDING @ STAIR RUN (SEE DET. C-B")

2x6 @ 16" O.C. MAX.

EXIST. ROUGH PLASTER WALL FIN.

HEAD & JAMB DET. SIMILAR TO DET. "A-G"

EXIST. SMOOTH PLASTER WAINSCOT

NEW WAINSCOT- MATCH SMOOTH TEXTURE @ EXIST.

NEW CHAIR RAIL TO MATCH EXIST

EXIST. CH. RAIL REMOVED

NEW BASE BOARD TO MATCH EXIST

EXIST. BANDED REMOVED AS SHOWN

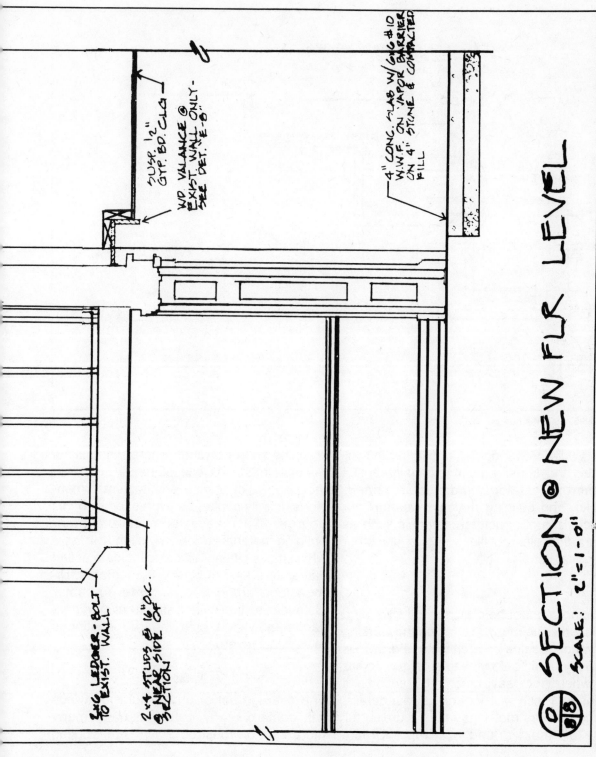

SECTION @ NEW FLR LEVEL

SCALE: 1" = 1'-0"

⊘ 15/8

2x6 LEDGER - BOLT TO EXIST. WALL

2x4 STUDS @ 16" O.C. @ NEAR SIDE OF SECTION

SUSP. 1/2" GYP. BD. CLG.

W.D. VALANCE @ EXIST. WALL ONLY - SEE DET. E-8"

4" CONC. SLAB W/ 6x6 #10 W.W.F. ON VAPOR BARRIER ON 4" STONE & COMPACTED FILL

Fig. 15-7. Another section of the stairs shown in Fig. 15-4—from yet a different angle.

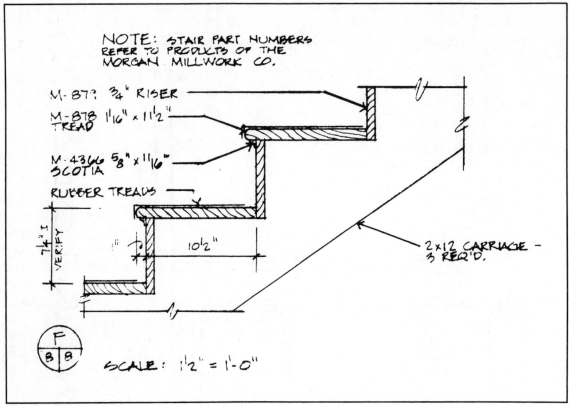

NOTE: STAIR PART NUMBERS REFER TO PRODUCTS OF THE MORGAN MILLWORK CO.

M-877 ¾" RISER

M-878 1¹⁄₁₆" x 11¹⁄₂" TREAD

M-4366 ⅝" x 11¹⁄₁₆" SCOTIA

RUBBER TREADS

7¹⁄₄" ± VERIFY

10¹⁄₂"

2 x 12 CARRIAGE - 3 REQ'D.

SCALE: 1¹⁄₂" = 1'-0"

Fig. 15-8. Stair tread detail.

While preparing the drawings, the architect and draftsmen will continually consult numerous catalogs, architectural handbooks, and building material literature for exact sizes of structural materials such as doors, floor tile, roof tile, windows, and other essential items.

Scale

Buildings cannot be drawn to full size, so a reduced scale must be used on the working drawings. Scales on architectural drawings will vary from floor plans to enlarged details anywhere from, say, 1 inch = 60 feet on plot plans to 1 inch = 1 foot on certain details. Items such as moldings may be drawn full size on the working drawings. Unlike machine shop drawings where dimensions are often made to close tolerance such as in thousandths of an inch, architectural dimensions usually cannot be expressed with this degree of accuracy. Lumber and other building materials often vary slightly in size, which makes it impractical to indicate actual values to edges of structural members. Dimensions are given as center-to-center distances so that even if structural members vary somewhat in size, the true location is always achieved.

Notes

The average set of architectural drawings will contain many notes—usually many more than on most other types of working drawings. The notes are used to augment dimensioning information or to explain some detail that otherwise cannot easily be shown clearly.

Symbols and Terms

Note from the examples shown in this chapter and others that many symbols are used in preparing architectural drawings. These symbols represent types of material, structural elements, or construction details. Most of these symbols are standardized and are familiar to all architects or workmen. If any variations are encountered, a legend or symbol list is provided to clarify the situation. Whenever new material or equipment is employed for which there are no standard symbols, the items are identified by num-

bered or lettered circles and described completely by note, detail, or in the specification form. Special symbols are also used to designate various structural, electrical, plumbing, and HVAC units.

The persons reading architectural drawings must have a good knowledge of common terms used in building construction. Many of these terms are included in the glossary.

Specifications

When the drawings are nearing completion for a given project, the architect usually begins work on the written specifications that generally spell out all construction details such as materials to be used, quality of workmanship, type of equipment, finishes,

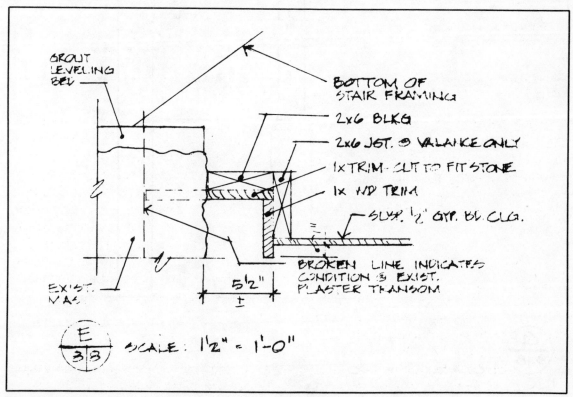

Fig. 15-19. Valance detail.

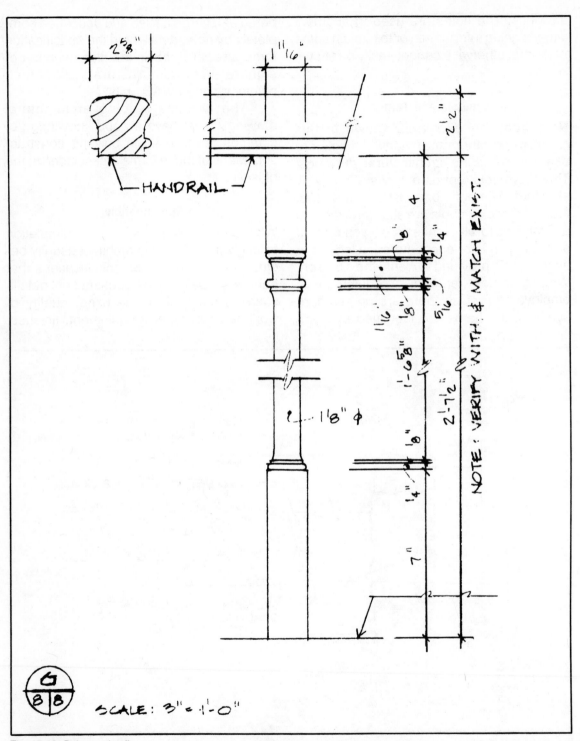

2⅝"

1¹¹⁄₁₆"

HANDRAIL

2½"

NOTE: VERIFY WITH & MATCH EXIST.

4

¹⁄₈"

¹⁄₄"

¹¹⁄₁₆"

⅛"

⁵⁄₁₆"

1'-6⅝"

2'-7½"

¹⁄₈"

¹⁄₄"

7"

1⅛" ⌀

Fig. 15-10. Baluster detail.

SCALE: 3" = 1'-0"

⁶⁄₈ ⁸⁄₈

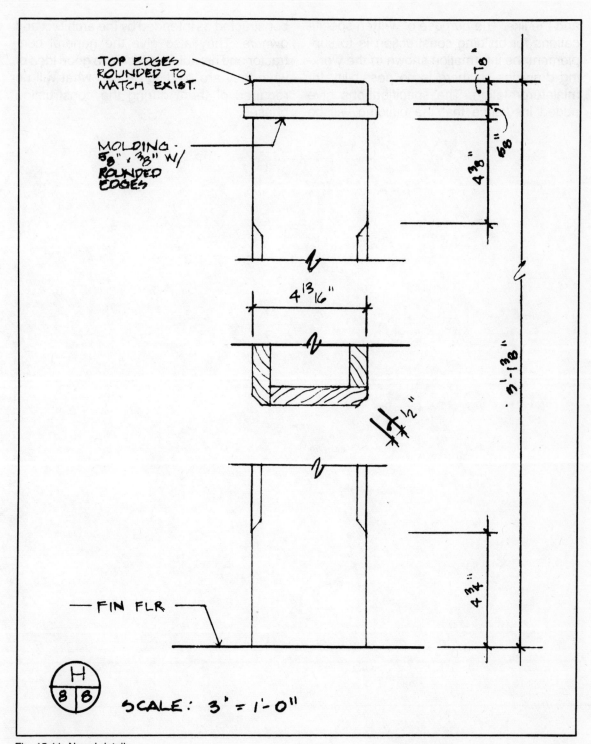

TOP EDGES ROUNDED TO MATCH EXIST.

MOLDING. 5/8" x 3/8" W/ ROUNDED EDGES

4 13/16"

1/2"

4 3/8"

6 5/8"

7/8"

3'-1 3/8"

4 3/4"

FIN FLR

⊕ H/8 8

SCALE: 3" = 1'-0"

Fig. 15-11. Newel detail.

and the like. The purpose of written specifications for building construction is to supplement the information shown in the working drawings so there is no possibility for misinterpretation. The specifications give added insurance that the building will be constructed as intended by the architect and owners. They also give the general contractor and his subcontractors a good idea of what they are bidding on and what will be required of them during the construction phases.

Chapter 16

Cabinet and Furniture Drawings

DRAWINGS USED IN THE CABINETMAKING INDUStry are similar to orthographic and pictorial drawings. The main advantage of a pictorial drawing is that it gives workmen a means of visualizing the finished product. The types of pictorial drawings most often found in the cabinetmaking trades include isometric, oblique, and perspective. In all three of these drawings an object is drawn in one view only—simulating a three-dimensional effect on a flat plane. The main disadvantages of all three types are that intricate parts cannot be pictured clearly, and all are difficult to dimension.

Of the three types mentioned, the oblique drawing is probably used more than any other—especially in exploded drawings (see Chapter 17). One face of the object is drawn in its true shape, and the other visible faces are shown by parallel lines drawn at the same angle (usually 45-30 degrees) with the horizontal. The lines drawn at a 30-degree angle are shortened and are therefore not drawn to scale.

Make an isometric drawing when you need a pictorial drawing drawn to scale. This type of drawing sometimes appears distorted as all isometric lines are drawn to scale. The edges are 120 degrees apart and are called the isometric axes. The three surfaces are called the isometric planes. The lines parallel to the isometric axes are called the isometric lines.

The receding lines are foreshortened one-half their actual length in a cabinet projection. The shortening of the receding lines tends to eliminate some of the distortion that is often quite noticeable in other projections.

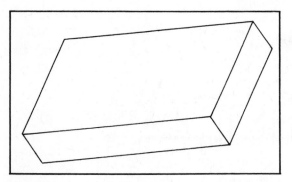

Fig. 16-1. Isometric drawing of a walnut base for a proposed jewelry box.

The receding axes may be drawn at any angle, but they are more commonly made at 30 or 45 degrees with the horizontal.

An isometric drawing of a walnut base for a proposed jewelry box is shown in Fig. 16-1. Note that all edges are drawn to scale, but the base looks somewhat distorted. In Fig. 16-2 this same base is drawn in a cabinet projection. While the latter is not drawn entirely to scale, it does give a more natural appearance to the base. While studying Fig. 16-2, note that the front width and length of the base are drawn to scale. The receding axes (to the front) are drawn at 30 degrees. While the thickness of the base is drawn to scale, the lengths of these lines have been shortened by one-half to give a more natural appearance.

An oblique projection may make any angle with the plane of projection. Receding lines vary from full to one-half scale, de-pending on the object being drawn. For all practical purposes, the angle of projection is kept between 30 and 60 degrees. The exact angle is chosen to better suit the situation at hand—that is, the shape of the object and the desired effects. The angle chosen should be the one giving the least amount of distortion. The object may also be placed in many positions—varying the receding lines—depending on what effect is desired.

When circles or arcs are to be made in cabinet drawings, templates are sometimes used. If none are available, the four-center ellipse system can be used to obtain perfect circles and arcs. This method is more time-consuming. When foreshortened receding axes are used in the drawing, the offset measuring system is used to draw circles. In this method a series of parallel reference lines are first located on a regular multiview projection as shown in Fig. 16-3. The curve or circle is located on these lines. The points are then transferred to similar parallel coor-dinates drawn on the cabinet view. The cir-

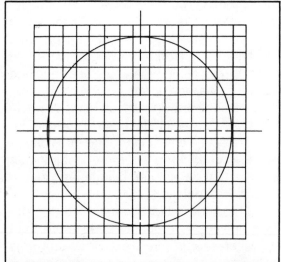

Fig. 6-3. The offset measuring system involves first drawing a series of parallel reference lines and then locating the curve or circle on them.

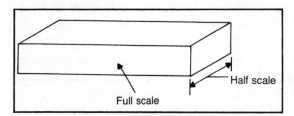

Fig. 16-2. Cabinet projection of the walnut base shown in Fig. 16-1.

Half scale

Full scale

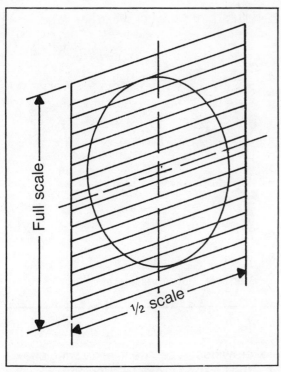

Fig. 16-4. Points of the circle drawn in Fig. 16-3 are transferred to similar parallel coordinates on the cabinet view.

cle or arc is completed with an irregular French curve (Fig. 16-4).

A cabinet drawing is dimensioned similar to an isometric drawing, but all dimensions are placed in the planes to which they apply. They are kept off the object as much as possible. When greater clarity is required, the dimensions are sometimes placed directly on the object.

AXONOMETRIC, DIMETRIC, AND TRIMETRIC PROJECTIONS

Another pictorial drawing is the *axonometric projection*. An object is represented by its perpendicular projection on a surface in such a way that a rectangular solid appears as inclined and shows three faces. An isometric drawing (Fig. 16-1) has angles of 30 degrees for its axes. In a *dimetric projec-*

tion two axes also make equal angles with a plane of projection. The third axis makes a different angle (Fig. 16-5). The edges that are parallel to the first two axes are foreshortened the same amount. The edges that are parallel to the third axis are foreshortened to a different value. An angle of 15 degrees was used for the two equal axes. In comparing Figs. 16-1 and 16-5, it is obvious that a dimetric projection is similar to an isometric projection except that the former has less distortion.

When reading dimetric drawings, remember that two separate measurement scales have been used. These drawings cannot be scaled like isometric projections.

Dimetric drawings are placed on the paper so that the object's main axis is at a 90-degree angle to the horizontal. The receding axes will be located at various angles depending on the desired effect and the shape of the object.

Another pictorial drawing is the *trimetric* projection. All three axes of the object make different angles with the plane of projection (Fig. 16-6). Lines parallel to each of the three axes have different ratios of foreshortening.

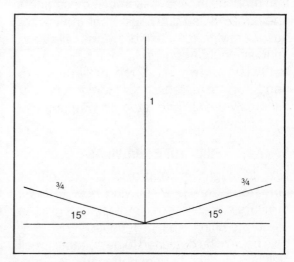

Fig. 16-5. A dimetric projection.

233

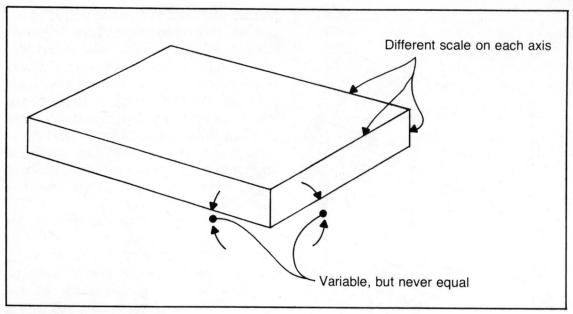

Different scale on each axis

Variable, but never equal

Fig. 16-6. A trimetric projection.

Three different trimetric scales must be used.

Although drawings resembling oblique drawings and made in a certain way are termed cabinet drawings, such drawing techniques are used in other ways besides for cabinets and furniture. Drawings used for the production and assembly of cabinets and furniture use drawings other than pictorial ones. Many of these drawings utilize the conventional orthographic method—that is, multiview drawings. Such drawings are necessary to establish all construction or assembly details and often accompany pictorial drawings.

FURNITURE DRAWINGS

How-to magazines or project books contain numerous drawings showing how to build furniture pieces from a simple footstool to an elaborate china closet. Because most projects have to be designed to be suitable for all skill levels, each project normally uses several drawings. A perspective drawing shows how the finished project will look as if viewed by the human eye. An exploded view shows essential parts and how they fit together.

Orthographic drawings or multiviews show all construction details and exact dimensions. Enlarged detail drawings show how certain construction details are to be done such as fitting a joint, the radius of a scroll, and the like. Written instructions and a materials list also enable the reasonably skilled do-it-yourselfer to build the project as the designer intended.

Many amateurs attempt the more complicated projects without first building the less difficult ones. The project usually ends up as kindling. The amateur builder is discouraged from ever attempting another home project. Even if you are very adept at reading specifications and blueprints, almost every woodworking project requires some skill that comes only with experience.

Learning to read drawings and specifications is a step in the right direction if you attempt do-it-yourself projects. You'll have to know how to safely use hand and power tools, but understanding every detail of how a particular item is built is just as important.

WORKBENCH

Let's look at a relatively easy-to-build work-bench. An exploded view is shown in Fig. 16-7. This drawing shows how each piece is cut and shaped and how the pieces fit to-gether. Each wood member is fully de-scribed along with some construction de-tails. An orthographic view showing how to cut the plywood from a 4 × 8 sheet is shown in Fig. 16-8. You should be able to build a bench closely resembling the one shown by following these instructions.

Lay out the diagram for the top pieces and shelf on a piece of 4-foot-by-8-foot-by-

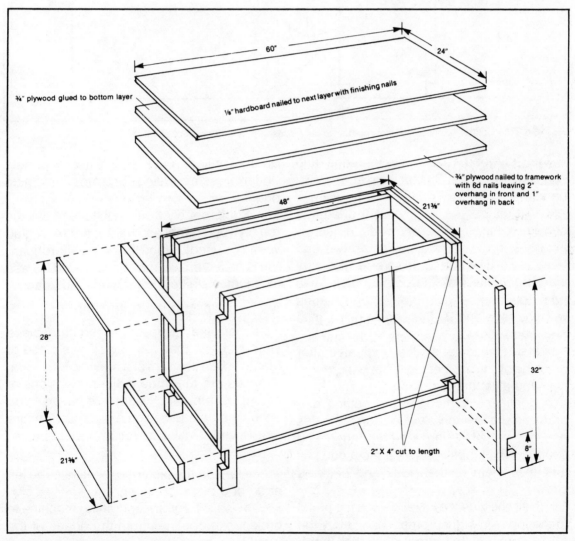

Fig. 16-7. Exploded view of a workbench.

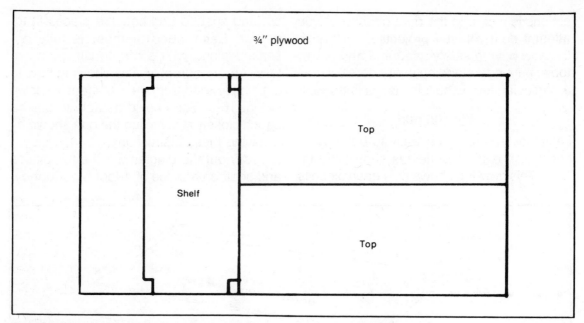

Fig. 16-8. Orthographic or plan view showing how to cut pieces from a 4×8 sheet of plywood.

¾-inch plywood. Use a straightedge and square for the lines. Cut out the pieces with a handsaw.

Cut all 2 × 4s and notch the legs as shown in Fig. 16-7. Mark and drill screw holes as indicated. Assemble the two end frames with glue and wood screws. You may want to countersink the screw holes for a neat appearance. Join the two end frames with the four long 2 × 4s, again using glue and wood screws.

Insert the lower shelf as indicated after notching all four corners to fit the bench legs. Nail and glue them in place.

Continue by marking the cutting diagram on the ⅜-inch plywood for the sides and back again using a straightedge and square. When these panels are cut out, nail and glue them to the sides and back as shown.

Nail and glue the lower ¾-inch top panel to the top rails of the frame, followed by the next panel and the sheet of ⅛-inch hard-board (nailed only). Keep these panels under pressure (using C-clamps) until glue dries.

You probably would not have to nail the top plywood panel and hardboard (only glue would be used). When the ⅛-inch hardboard top is nailed only, it may be removed in case it becomes stained, scratched, or damaged.

DESIGNING FURNITURE

You may like to design a furniture piece similar to one that you saw. If this piece of furniture is in the form of a drawing or sketch, chances are the three major dimensions of height, depth, and width are given. You should be able to design a piece of furniture that is very similar by following this procedure.

●Fasten the picture or sketch to the left edge of a piece of paper.

●Use a T square or drafting machine to project straight lines from the edges of the furniture piece.

●Suppose that the overall height is 30 inches. Select a scale in which 30 units equal in length will fit diagonally between the two horizontal lines. Place the first division point on the scale on the top line and the last division point on the bottom line.

●Project lines until they intersect the diagonal line for each dimension needed. It is then easy to count the number of units or inches for this part. The same method can be used to find other dimensions.

When building wood projects or looking over the drawing for a furniture project, matters are helped if you have some idea of what the designer had in mind when the project was in the design stage. The following guidelines will help you approach the subject of blueprint reading more intelligently.

During the preliminary steps of furniture design, the need and the exact use of the piece must be determined along with the style. A sound product ideally has an attractive appearance and fits into the space available. It should also have a distinctive style.

Develop several rough sketches of the piece in question. When a particular furniture style has been decided on, you should become thoroughly familiar with the style's characteristics. If an Early American style is desired, you should know that most corners are rounded rather than joined at a sharp edge. If the piece has drawers, they are of the lip type—not flush as found on most contemporary pieces.

You can experiment and may even build models of the proposed piece of furniture. Cardboard or balsa wood is a good choice for constructing a model of a piece of furniture. The model and sketches are then developed into a final drawing. A bill of materials should also be included.

When reviewing various drawings of furniture, judge the product to see if it fits your needs and is satisfactory. Good design involves sensitivity to beauty. As you study more drawings of furniture, you will learn more about design. You will become more sensitive to it and will be able to detect poor designs and those that do not fit your needs.

Chapter 17

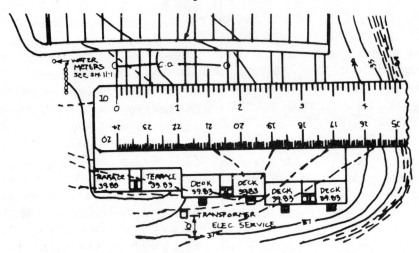

Exploded Views

EXPLODED VIEWS ARE EXCELLENT WAYS OF showing how an object is to be put together. Pick up any project or how-to magazine and you'll find exploded views used extensively to better describe how a project goes together.

An exploded view is a pictorial drawing of an object that shows the parts separated but in correct relationship to each other. Let's look at an exploded view of a workbench (Fig. 17-1). This is an oblique drawing with only portions of the bench exploded. There should be little question as to how this bench is constructed. Even with no dimensions, an experienced workman should be able to build a similar bench. If the bench is to be built to specified measurements or dimensions, you would simply number the various parts and then list the dimensions of each part in a schedule. This exploded view could be supplemented with orthographic drawings. The dimensions would be directly on the plans.

Exploded views come under pictorial drawings in that they use either isometric, oblique, or perspective drawings for the various parts of the object to be exploded. With several pictorial drawings of the various parts of an object in a single view, you can see how the individual parts of a mechanism fit together. These drawings are particularly valuable to those who are unable to read multiview drawings. Exploded drawings range from very simple designs to elaborate shaded illustrations. Perspective drawings are most often used in an exploded drawing because they give more pleasing appearance to the drawing.

To better understand how to read an exploded view, let's review how such a drawing is made. Usually a perspective of the entire assembly is done. A piece of trac-

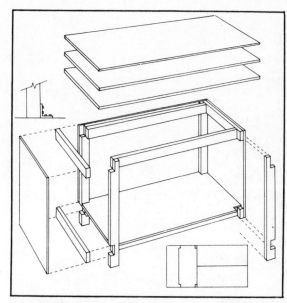
Fig. 17-1. Exploded view of a workbench showing all wood members and how they relate to each other.

ing paper is placed over the assembly drawing, and one or more parts are drawn with the perspective axes clearly shown. The sheet of tracing paper is then removed, and another sheet is positioned on which some other part is drawn. When all the parts have been drawn, the individual sheets are arranged so each part is in its normal spread-out position. A final drawing is then made. The finished drawing will show the various parts of an assembly in a separate or pulled-out position. All the parts are in the correct order for reassembly.

AMMUNITION RELOADING TOOL

An exploded view of an ammunition reloading tool manufactured by Dillon Precision, Inc., is shown in Fig. 17-2. This drawing aids in repairing the tool and ordering replacement parts. The drawing will aid you in disassembling the tool, replacing the broken part, and then reassembling the tool. To positively identify the part in question, each

one is numbered to correspond with the schedule in the left-hand side of the drawing.

Let's assume that you lost coil spring number 30 in Fig. 17-2 and want to replace it with the correct part. You would find the number RL 450 30 in the parts list. This part is called primer slide return spring. Order the part by name and reference number from the manufacturer.

After you receive the part, refer to Fig. 17-2 to see how the part fits into the tool. One end of the spring fits into a hole in the frame (number 1). Although it cannot be seen in Fig. 17-2, we will assume that some type of stud is provided in this hole on which to hook the "eye" of the spring. Because this spring will cause bar number 29 to spring back when pulled outward, the other end of the spring will connect to pin number 31, which is the primer slide return spring post. Note the ring groove in the bottom of the post that prevents the spring eye from slipping off. If there is a question as to where a certain part may go on the tool, broken lines are used to "guide" the part to its proper place.

Exploded views are frequently used by firearm manufacturers to illustrate how different models are assembled and disassembled. Exploded drawings of firearms also allow you to compare the actual parts in the gun so that none are missing or broken. The drawings act as troubleshooting charts.

WINCHESTER MODEL 1895 RIFLE

Exploded views were not always available to gun owners and gunsmiths. Manufacturers used to merely provide instructions on how a particular firearm was assembled and disassembled. Some parts were drawn in a perspective view, but few were arranged in their proper sequence. At one time the famous Winchester Repeating Arms Com-

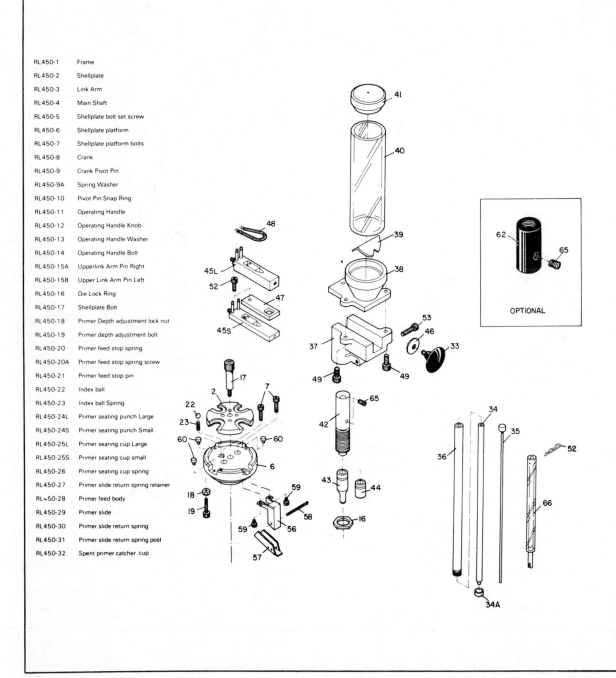

RL450-1	Frame
RL450-2	Shellplate
RL450-3	Link Arm
RL450-4	Main Shaft
RL450-5	Shellplate bolt set screw
RL450-6	Shellplate platform
RL450-7	Shellplate platform bolts
RL450-8	Crank
RL450-9	Crank Pivot Pin
RL450-9A	Spring Washer
RL450-10	Pivot Pin Snap Ring
RL450-11	Operating Handle
RL450-12	Operating Handle Knob
RL450-13	Operating Handle Washer
RL450-14	Operating Handle Bolt
RL450-15A	Upperlink Arm Pin Right
RL450-15B	Upper Link Arm Pin Left
RL450-16	Die Lock Ring
RL450-17	Shellplate Bolt
RL450-18	Primer Depth adjustment lock nut
RL450-19	Primer depth adjustment bolt
RL450-20	Primer feed stop spring
RL450-20A	Primer feed stop spring screw
RL450-21	Primer feed stop pin
RL450-22	Index ball
RL450-23	Index ball Spring
RL450-24L	Primer seating punch Large
RL450-24S	Primer seating punch Small
RL450-25L	Primer searing cup Large
RL450-25S	Primer seating cup small
RL450-26	Primer seating cup spring
RL450-27	Primer slide return spring retainer
RL450-28	Primer feed body
RL450-29	Primer slide
RL450-30	Primer slide return spring
RL450-31	Primer slide return spring post
RL450-32	Spent primer catcher cup

Fig. 17-2. Exploded view of an ammunition reloading tool listing part numbers to aid in ordering replacement parts and also in assembly and disassembly (courtesy Dillon Precision, Inc.).

240

RL450-33	Primer slide knob/Powder bar knob
RL450-34S	Primer magazine Small
RL450-34L	Primer magazine Large
RL450-34A	Primer magazine lock ring
RL450-35	Primer magazine follower
RL450-36	Primer magazine shield
RL450-37	Powder measure base
RL450-38	Powder measure top
RL450-39	Powder measure baffle
RL450-40	Powder measure tube
RL459-41	Powder measure lid
RL450-42	Powder die
RL450-43	Powder funnel/pistol expander
RL450-44	Rifle powder funnel
RL450-45L	Powder bar Large
RL450-45S	Powder bar Small
RL450-46	Powder bar stop
RL450-47	Powder bar spacer
RL450-48	Powder bar spring
RL450-49	10 by 24 by 3/4 inch cap screw
RL450-50	10 by 24 by 3/8 inch set screw
RL450-51	Primer slide stop nut
RL450-52	10 by 24 by 5/8 cap screw
RL450-53	Powder clamp bolt
RL450-54	
RL450-55	Spent primer cup screw
RL450-56	Spent primer catcher bracket
RL450-57	Spent primer catcher chute
RL450-58	Spent primer catcher pin
RL450-59	Spent primer catcher screws
RL450-60	Locator buttons (3)
RL450-61	
RL450-62	Powder measure adaptor
RL450-63	Main shaft pivot pin
RL450-64	Main shaft pivot pin retaining screw
RL450-65	Primer slide/Powder die set screws (10-32 by ¼")
RL450-66L	Primer pick up tube Large
RL450-66S	Primer pick up tube Small

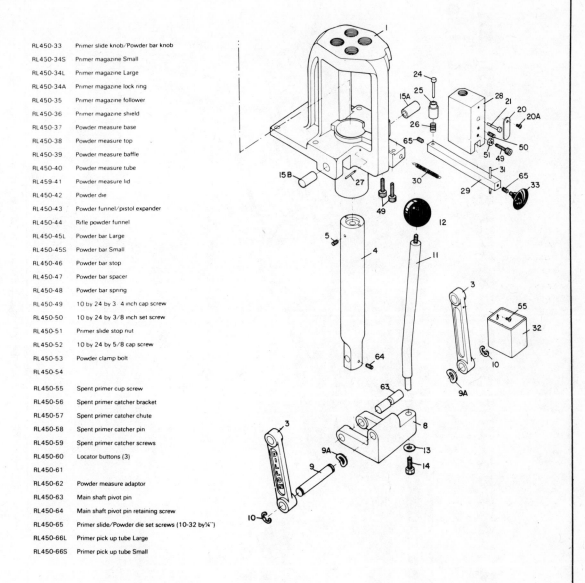

THE LATEST WINCHESTER REPEATING RIFLE.
MODEL 1895.

.30 U. S. Army.
.38-72 W. C. F.
.40-72 W. C. F.

TO DISMOUNT THE ARM.

Remove the fore-arm, take out the carrier spring and take off the butt stock.   Open the lever and remove the carrier screw and magazine tip screw.   These are the two lower screws on the forward end of the frame.   Remove the magazine with inclosed carrier, turn out the mainspring strain screw, take out the mainspring screw and mainspring.   Close the lever and take out the hammer screw and hammer.   Remove the finger-lever pin stop screw and drive out the finger-lever pin from the forward hole in the right side of the frame.   Remove the link pin.   (This connects the link and the lower tang.)   The finger-lever, link and trigger can then be removed together.   Take out the sear spring screw—the forward screw on the bottom of the tang—and the sear pin.   Take out the locking bolt and breech-bolt.

Fig. 17-3. Sectional view of a Winchester Model 95 rifle as pictured in an old catalog (courtesy Winchester Repeating Arms Company).

COMPONENT PARTS
OF THE
WINCHESTER REPEATING RIFLE,
MODEL 1895.

PRICE LIST OF COMPONENT PARTS.

Assembling Screw,	$0.05	Finger Lever Pin,	$0.10
Barrel, Round, Nickeled Steel,		Finger Lever Link Split Pin,	.20
.30 Calibers,	10.00	Friction Stud,	.10
Barrel, Round, .38 and .40 Calibers,	6.25	Friction Stud Spring,	.05
" Octagon, .38 and .40 Calibers,	8.00	Friction Stud Pin,	.02
Breech Bolt,	3.00	Firing Pin,	.50
Butt Stock,	1.40	Firing Pin Lock,	.35
Butt Plate,	.80	Firing Pin Lock Pin,	.03
Butt Plate Screws (2), each,	.05	Hammer,	.50
Carrier,	1.00	Hammer Screw,	.10
Carrier Screw,	.05	Link,	1.00
Carrier Spring,	.25	Link Pin,	.05
Carrier Stop Pin,	.03	Locking Bolt,	1.50
Carrier Cradle,	.40	Mainspring,	.30
Carrier Cradle Pin,	.03	Mainspring Screw,	.05
Carrier Cam Lever,	.50	Mainspring Strain Screw,	.05
Carrier Cam Lever Pin,	.03	Magazine,	4.00
Ejector,	.30	Magazine Tip Screw,	.05
Ejector Pin,	.05	Receiver,	8.00
Ejector Spring,	.10	Sear,	.40
Extractor,	.25	Sear Pin,	.03
Extractor Pin,	.03	Sear Spring,	.20
Fore-Arm,	1.00	Sear Spring Screw,	.05
Fore-Arm Stud,	.15	Trigger,	.30
Fore-Arm Screw,	.05	Trigger Pin,	.03
Finger Lever,	1.50	Tang Screw,	.10

Fig. 17-4. Perspective drawings of individual parts for the rifle shown in Fig. 17-3 (courtesy Winchester Repeating Arms Company).

pany only showed a sectional view of their arms—like the one of a Model 1895 in Fig. 17-3. Written instructions accompanied the drawing.

The instructions accompanying the gun also had a parts list and some drawings of the individual parts as shown in Fig. 17-4. With these written instructions, a sectional view, and drawings of individual parts, the experienced gunsmith seldom had any difficulty in assembling the gun. Some owners who might want to completely disassemble their gun for a thorough cleaning might not find the task so easy. Most gunsmithing business is brought about by amateurs trying to completely disassemble their own guns. Most of them have little trouble getting

the gun apart, but getting it back together again is a different matter. Manufacturers started providing exploded views of their weapons such as the one shown in Fig. 17-5. With a parts list and each part identified, the job of disassembling and identifying defective parts was made much easier.

Assembly and disassembly instructions accompany the exploded view. Where written instructions may be unclear, supplemental pictorial drawings such as those in Fig. 17-6 are also provided.

VALUE OF EXPLODED VIEW DRAWINGS

Exploded views and pictorial drawings are invaluable to manufacturers of products that

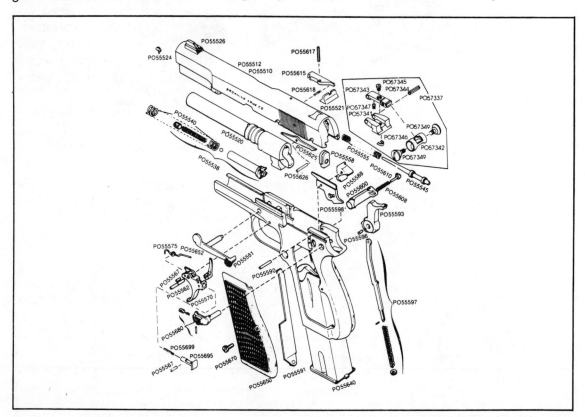

Fig. 17-5. An exploded view of modern firearms is almost compulsory for proper assembly and disassembly. Such drawings are also excellent for troubleshooting broken or missing parts.

must or are likely to be disassembled by users. Besides providing the users with helpful data, the drawings and written instructions save the manufacturers much money each year. How? Imagine the number of letters and phone calls one manufacturer would get each year if such drawings and instructions were not provided. Each manufacturer would have to hire several employees just to answer the phone and to type replies to users' letters. The amount of money spent for the drawing and printed instructions is well worth it. Such helpful information doesn't hurt the manufacturer's reputation either.

Most readers will probably come across exploded views more frequently in plans for how-to projects than any other time. Several magazines contain useful projects designed to be built by homeowners and do-it-yourselfers. Most projects will have at least one exploded view of the project to show how each part relates to the other. Many of

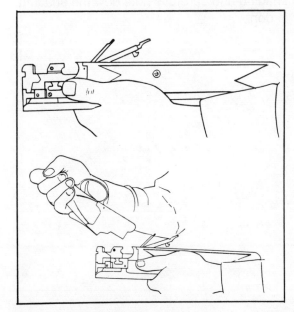

Fig. 17-6. Supplemental pictorial drawings are sometimes used along with exploded views.

the projects would never get built without these exploded views.

Book publishers also rely heavily on exploded views when describing how to build a project. These drawings are not used as frequently as they should be. One reason is the lack of draftsmen or artists to do the drawings. Any experienced draftsman should be capable of making an orthographic view of almost any object provided enough information is provided. Few people can draw an effective exploded view. They must know how the object to be drawn is assembled and then have the talent to illustrate the details.

MAKING AN EXPLODED DRAWING

To better understand how to read an exploded view, let's look at how one is made. Remember that exploded views may vary from simple sketches to elaborate shaded illustrations that are almost like photographs of an object.

Exploded drawings can be prepared either in axonometric, oblique, or perspective form. As mentioned previously, the perspective technique is preferred by most manufacturers because it gives a natural appearance of the object in question. When parts are strung out along a single axis by the axonometric or oblique method, some distortion is inevitable.

To make an exploded drawing, a perspective of the object is first obtained. This may be a photograph of the object placed in the position the drawing is to be made, but more than likely the perspective will be a drawing of how the object will appear when it is assembled or completely constructed. A piece of tracing paper is next placed over the perspective, and one or more parts are drawn with the perspective axes clearly shown. The overlay sheet is

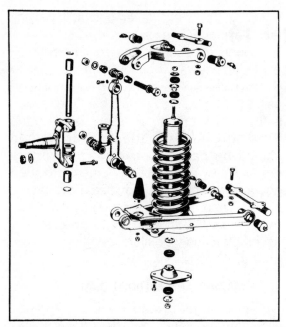

Fig. 17-7. Shading gives a more natural appearance to any drawing.

ranged so each part is in its normal spread-out position. A final drawing is then made.

An experienced illustrator will use the same sheet of drawing paper rather than several sheets. He first draws the main part of the object and then moves the overlay or tracing paper the correct distance, and in the correct relationship to the main part, and draws another part. The sheet is moved again, another part is drawn, and so on, until all the parts are drawn in their proper location.

At this point it may be necessary to make the exploded drawing appear more natural—to bridge the gap between an actual photograph and the natural object. Shading is often used to make the drawing appear more natural (Fig. 17-7). Shading is simply a technique of varying the light intensity on the surfaces by lines or tones. A complete book could be written on this subject alone, so no more space will be spent on it. You should review several drawings using shading and carefully note how the shading is done.

removed, and another sheet is positioned on which some other part is drawn. This process continues until all the parts have been drawn. Then the individual sheets are ar-

Chapter 18

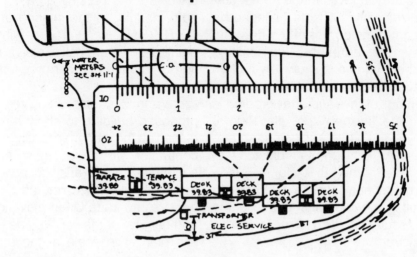

Site Plans

A *SITE PLAN* IS A PLAN VIEW (AS IF VIEWED FROM AN airplane) that shows the property boundaries and the building(s) drawn to scale and in its (their) proper location(s) on the lot. These plans will also include (where applicable) sidewalks, drives, streets, and similar details. Utilities such as water lines, sanitary sewer lines, telephone lines, and electrical power lines must also appear on site plans—sometimes on the original site plan furnished by the architect and other times on a separate site plan prepared by the engineering firm.

The initial site plan will be prepared by a certified land surveyor from information obtained from a deed description of the property. This property survey will show only the property lines and their lengths as if the property was perfectly flat. If additional information is necessary, a complete field or topographic survey must be made. This top-

ographic survey, in addition to indicating property lines, will also show the physical characteristics of the land by using contour lines, notes, and symbols. The physical characteristics may include: all property lines, pertinent landmarks, the direction of the land slope, and whether the land is flat, hilly, wooded, swampy, high, low, and other physical features.

Every homeowner will have a description of his or her property recorded in the deed files of the county clerk's office or other legal office, depending on which section of the country is involved. A site plan made of your lot by a certified land surveyor may also be needed. You should have several copies of this drawing made and keep them in your possession. They will help you settle land disputes, plan landscape arrangements, route utility lines, and the like. You should also know how to read a site or plot plan.

PLOT PLAN OF A BRANCH BANK

Figure 18-1 is a plot plan of a branch bank project. This drawing was made from a surveyor's site plan of the lot. Note the legend in the upper right-hand corner of the drawing. The first symbol represents the existing contour of the land and is shown by broken lines. The solid lines represent the proposed finished contour of the land after grading is completed. The finished spot elevation is indicated by a rectangle enclosing the elevation in feet. A plus sign (+) indicates feet above sea level. A negative sign (−) indicates feet below sea level. A 1935 bench mark located in the southeast corner of the property is assumed to be 100 feet-0 inches above sea level.

The shaded area on the site plan indicates the proposed asphalt paving for the driveway and parking areas. You will also note marking in the shaded area such as "OUT," "IN," etc.; these markings are to be painted on the asphalt to alert traffic to direction of travel and other precautionary measures.

Various angles are given around the proposed building to show how it is to be placed on the lot. When footings are dug, the contractor will "shoot" these angles with a surveyor's level/transit to assure that the building will be located as the architect/owners intended. Other elevation markings and general notes can also be found on the site plan. The scale at the bottom of the

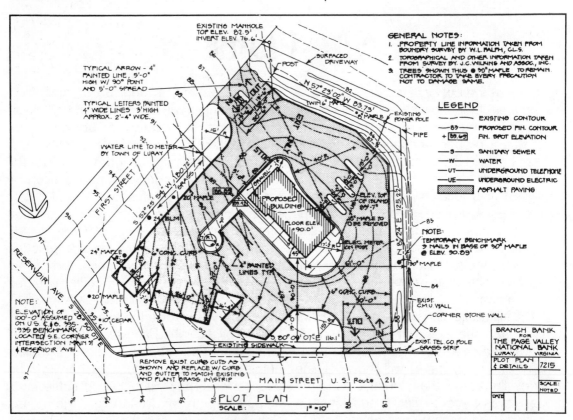

Fig. 18-1. Plot plan of a branch bank. This site plan was included with a set of architectural drawings to indicate the location of the building on the lot, the amount and type of grading required, and other necessary information.

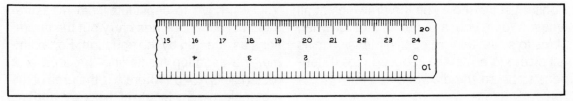

Fig. 18-2. Several graduations of engineer's scales.

ENGINEER'S SCALE

drawing, under the drawing title, tells the person reading the drawings that 1 inch on the drawing is equal to 10 feet on the site. The reader should study this plan until he or she is thoroughly familiar with all details described on it.

Nearly all site plans are drawn to some scale, and the *engineer's scale* is the type normally used. The engineer's scale is similar to the architect's scale discussed in Chapter 5, except that an inch is divided into 10, 20, 30, 40, 50, or 60 equal parts. Some engineer's scales are shown in Fig. 18-2. The scale marked 10 is read 1 inch = 10 feet

or some other integral power of 10 such as 1 inch = 100 feet, 1 inch = 1000 feet, and so on. Because the engineer's scale is the chief means of making scaled site plans, its use should be thoroughly understood by those who must read site plans.

The engineer's scale is used by placing it on the drawing with the working edge away from the user (Fig. 18-3). The scale is then aligned in the direction of the required measurement. The dimension is read by looking down over the scale.

Although the drawing itself may appear reduced in scale, depending on the size of the object and the size of the drawing sheet to be used, the actual true-length dimen-

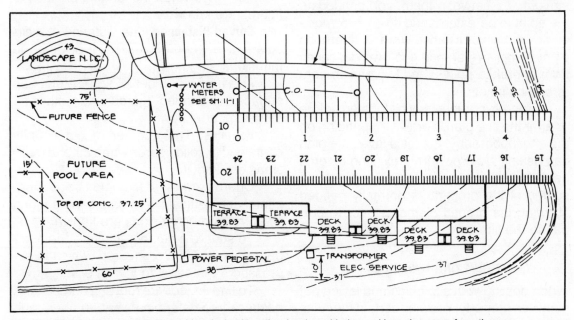

Fig. 18-3. The engineer's scale is used by placing it on the drawing with the working edge away from the user.

sions must be shown on the drawings at all times. When you are reading or drawing plans to scale, think of each dimension in its full size and not in the reduced size it happens to be on the drawing.

USING SITE PLANS

The first step is to locate a point. You must know where it should be placed in relation to known points. This is similar to giving directions to someone. You must first start them off along a street that they know or from a building that they know, and so forth. It would be impossible to give directions without some reference point. The same is true when reading or using site plans.

In most cases you must begin with two known points when laying out property lines. You can then join these points with a straight line that serves as your reference line. Then you can easily establish additional reference points along the reference line. When you have your two reference points and the straight line between them, you can locate the point to be established if you know: one angle and one distance from a reference point along a reference line, two distances from reference points, and two angles from reference points.

Property corners may be located before or after the site plan is drawn, depending on the site topography. If a lot is level, the plan can usually be drawn from existing records describing the property. For hilly property, a survey of the property is normally needed showing the ground elements prior to preparing the site plan.

SURVEYING FORMULAS

When dealing with site plans and laying out points on property, knowing certain formulas and calculations will be extremely helpful.

You will have to substitute field measurements in equations and carry out the operations as directed by the equation. For example, the equation for finding the area of a rectangle is $A = bh$ where A is the area of the rectangle, b is the length of the base, and h is the height or width. If the base and height are known, simply substitute the known quantities in the equation and carry out the mathematical operations. In the area equation multiply the base times the height, and the answer is the area. If the base is 30 feet and the height is 20 feet, the area is 600 feet ($20 \times 30 = 600$).

In dealing with site plans you will be using equations to help calculate the following.

Perimeters. The total distance around figures when certain dimensions are given.

Sides of Figures. The length of the sides of a rectangle, triangle, etc., when you know an angle and other sides of the figure.

Angles of Figures. The angle in a triangle or other figure when the lengths of sides of figures are known.

Curved Distances. The distance of a circular curve when angles, radii, or other dimensions are known.

Areas. The number of square units on surfaces of bonded figures when other dimensions are known.

Volumes. The number of cubic units contained in solid figures when other dimensions are known.

Chords, tangents, angles, and horizontal and vertical curve data can be found when related dimensions are known.

To square a number, multiply the number by itself. Thus 5 squared, which is expressed 5^2, is equal to 5×5, or 25.

To find the *square root* of a number, it is necessary to determine what number, multiplied by itself, will give you the original number. To find the square root of 36, you must determine what number, multiplied by itself, will give you 36. The square root of 36 is therefore 6.

To indicate square root, the radical sign $\sqrt{}$ is used. Thus, $\sqrt{36}$ means the square root of 36 or the number 6.

In the square root example just used, it is a perfect square—one that did not have a remainder. Few numbers are perfect squares. Methods used to determine square root in the past have included the slide rule, using square root tables, using logarithms and log tables, and computing the square root. A pocket calculator can figure the square root of any number quickly.

The *perimeter* is the distance around the outside of an object or figure. The result will come out in basic units such as feet, yards, miles, etc. The perimeter is found by totaling the dimensions of the sides of a figure.

The following equations may be used to find the circumference or perimeter of a circle:

$$c = \pi d \text{ (or) } C = \pi(2r)$$

Area is the number of unit squares equal in measure to the surface. The area of a surface generally is found by multiplying the base dimension times the height dimension. Area measurements will always be in square units—square feet, square yards, square miles, and so forth.

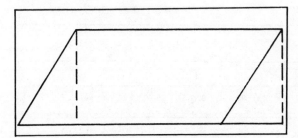

Fig. 18-4. Finding the area of the parallelogram.

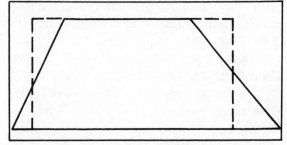

Fig. 18-5. The area of a trapezoid can be found by comparing it to a rectangle.

The area of a parallelogram is found using the same equation as for a rectangle, except that the height of a parallelogram is not the same dimension as any of its sides as shown in Fig. 18-4. Height refers to the distance of a line that is perpendicular to the base—not the slanted sides.

A trapezoid can also be compared to a rectangle. When it is, as in Fig. 18-5, the formula or equation to use is:

$$A \times \left(\frac{b1 + b2}{2}\right) h$$

The equation for finding the area of a triangle is similar to the one used for finding the area of a rectangle except that you divide the results by two. A triangle may be considered the same as a rectangle cut in half (Fig. 18-6).

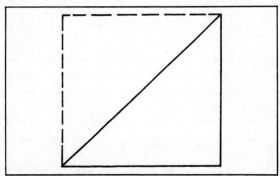

Fig. 18-6. As far as area is concerned, you may compare a triangle to one-half of the rectangle.

To find the area of a circle, square the radius and multiply the result by pi (3.14159).

These equations should be helpful in determining the area of a plot or site plan, which is usually referred to as acres or parts of an acre.

OTHER SCALES

The *mechanical engineer's scale* in Fig. 18-7 is typical of those currently in use. The major unit on these scales usually represents 1 inch, and the subdivisions represent the commonly used fractions of an inch—usually in multiples of ½ such as ½, ¼, ⅛, 1/16, etc.

The *decimal scale* shown in Fig. 18-8 is designed with decimal dimensioning rather than fractional. Two-place decimals are normally used. They are made in even numbers, so a two-place decimal results when halved.

The graduations on a *civil engineer's scale* represent decimal units. There are also scales in the metric system of measuring. These scales will more than likely eventually replace the inch, fractions, and decimals of an inch.

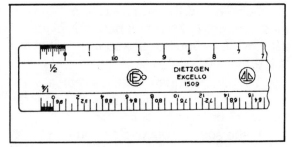

Fig. 18-7. Typical mechanical engineer's scale.

All scales are either open divided or fully divided. Open divided scales are those on which the main units are numbered along the whole length of the edge, with an extra unit fully subdivided in the opposite direction from the zero point. The subdivided unit shows the fractional graduations of the main unit. Open divided scales often have two complete measuring systems on one face. For example, ¼-inch and ⅛-inch scales may appear on one face of the architect's scale (Fig. 18-9).

Fully divided scales have all the subdivisions along the entire length of the ruler so that several values from the same origin can be read without having to reset the scale. This is accomplished by double numbering, either to permit both right-to-left and

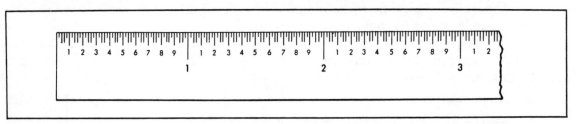

Fig. 18-8. A decimal scale is used for decimal dimensioning.

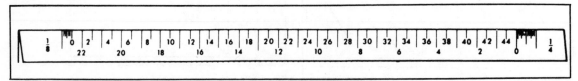

Fig. 18-9. Open divided scales often have two complete measuring systems on one face.

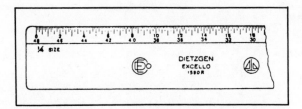

Fig. 18-10. Fully divided scales have all the subdivisions along the entire length of the ruler so that several values from the same origin can be read without having to reset the scale.

left-to-right readings, or to provide two different scales on one face (Fig. 18-10).

ABBREVIATIONS AND DEFINITIONS

The following abbreviations and definitions are those most used on plot and site plans and surveying drawings. Note the symbols in Fig. 18-11.

木	Instrument or instrument man	∠	Right angle
N	Notekeeper or recorder	⌁	Same both sides of line
φ	Rod or rod man		(usually ownership)
O	Iron pipe	@	AT
□	Stake	---------	Centerline or traverse line
⊥	Plane table setup	℞	Property line
⊡	Stadia station	=	Equal
⊙	Traverse station	≡	Identically equal to or exactly
■	Bench mark	≠	is not equal
<	Angle point or P.I.	~	Similar to
∡	Angle	≈	Approximately equal
#	Number or pound	≅	Is congruent with
Δ	Delta, central angle	<	Is less than
Δ /2 Δ /4	½ delta, ¼ delta, etc.	>	Is greater than
φ	Phi, latitude	≥	Equal to or greater than
λ	Lambda, longitude	≤	Equal to or less than
ꝋ	Throw over	∴	Therefore
π	Pi = 3.14159265+	∵	Because
⊿¹	Slope 2 to 1	III ≀≀≀ III	Stream or river
Σ	Summation	⌢⌣	Intermittent stream
⊥	Perpendicular to	- - - - -	Drive or walk
▦	Building	～～～	Stone fence
/ /	Parallel to	△	Triangulation station

Fig. 18-11. Important symbols.

A	Area
Ac.	Acres
A. C.	Accounts collectible or asphalt concrete
A.C.B.	Asphalt concrete base
A. C. Pav.	Asphaltic concrete pavement
A. C. W. S.	Asphalt concrete wearing surface
Adj.	Adjusted

App.	Approximate
Asph.	Asphalt
Astr.	Astronomical
Ave.	Avenue
Az.	Azimuth
& or ⊄	And
B. B.	Brass bolt
B. C.	Beginning of curve
Bdry.	Boundary
Bk.	Book
B. L.	Building line
Bldg.	Building
Blk.	Block
Blvd.	Boulevard
B. M.	Bench mark
B. S.	Backsight
B. V. C.	Beginning vertical curve
C.	Cut
Calc.	Calculate
Cb.	Curb
C. B.	Catch basin
C. C. P.	Centrifugal concrete pipe
C. E.	City engineer, civil engineer, city engineer deed book
Cem.	Cement
C. F.	Curb face
Ch.	Chain
Chd.	Chord
Ch. X	Chiseled cross
C. I. P.	Cast-iron pipe
Ck.	Check
Ck. Elev.	Check elevation
C/L or ⊄	Centerline
C. M. P.	Corrugated metal pipe
Co.	County
Conc.	Concrete
Const.	Construction
Cor.	Corner
Cor. I. P.	Corrugated iron pipe

Cor. M. P.	Corrugated metal pipe
C. S.	County surveyor
Ct.	Court
C.T.	Copper tack
Ctr.	Center
Ctr. Ret.	Center return
C. W. S.	County wall sheet
D	Degree of curvature
Δ	Delta (central angle of curve or triangle)
Decl.	Declination
Def.	Deflection
Deg. (°)	Degree (3°, etc. when preceded by a figure)
Dest.	Destroyed
D. F.	Douglas fir
D. G.	Disintegrated granite
D. H.	Drill hole
Dia.	Diameter
Dist.	District or distance
D. M.	District map
D. M. H.	Drop manhole
Dr.	Drive
Drn.	Drain
E.	East
E.C.	End of curve
El. or Elev.	Elevation
Ellip.	Elliptical
Engr.	Engineer(ing)
E/O	East of
Eq.	Equation
E. R.	End of return
Esmt.	Easement
Etc.	Et cetera (and so forth)
E. V. C.	End vertical curve
Ex.	Existing
Ext.	External
F.	Fill
F. B.	Field book
Fd.	Found
F. H.	Fire hydrant
Fig.	Figure

F. L.	Flow line
F. S.	Foresight or finished surface
Ft.	Foot or feet
F. T.	Flush tank
G.	Gutter
Galv.	Galvanized
G. C.	Grade change
G. L.	Ground line
Gr.	Grade
Grd.	Ground
H. C.	House connection or head chainman
H. I.	Height of instrument
Hor.	Horizontal
Hy.	Highway
i.e.	That is
In.	Inch
Inst.	Instrument
I. P.	Iron pipe
J. C.	Junction chamber
J.S.	Junction structure
L.	Length of arc
Lat.	Latitude
L. Chd.	Long chord
L. D.	Local depression
L. & T.	Lead and tack
L. H.	Lamp hole
Lin. Ft.	Lineal foot or feet
Lks.	Links
Loc.	Locate or location
Long.	Longitude or longitudinal
L. S.	Land surveyor
Lt.	Left
M.	Map or meter
Max.	Maximum
M. B.	Map book
M. C.	Middle of curve

Meas.	Measure
Mer.	Meridian
Mi.	Mile
Mid.	Midway or middle
Mil. Rs.	Military reservation
Misc.	Miscellaneous
M.H.	Manhole
Min.	Minute (when preceded by a figure)
M. H. T.	Mean high tide
M. L. L. W.	Mean lower low water
M. O.	Middle ordinate
Mon.	Monument or monolithic
M. R.	Miscellaneous record
M. S. L.	Mean sea level
N.	North
N. & T.	Nail and tin
N. E.	Northeast
No.	Number
N/O	North of
Norm.	Normal
N.W.	Northwest
Obs.	Observe
O. D.	Outside diameter
O. R.	Official record
O. & W.	Opening and widening
Ord.	Ordinance
Orig.	Original
p. or pg.	Page
pp. or pgs.	Pages
Par.	Paragraph or parallel
Pav.	Paving
P. C.	Party chief
P. C. C.	Point of compound curve or Portland cement concrete
Perp.	Perpendicular
P. I.	Point of intersection
P. L.	Property line
Pla.	Place
P. L. P.	Property line produced

P. O. B.	Point of beginning
P. O. C.	Point on curve
P. O. T.	Point on tangent
P. P.	Power pole
P. R. C.	Point of reverse curve
Prod.	Produced
Prof.	Profile
Prop.	Proposed
pt.	Point
P. T.	Point of tangent
Pvt. R/W	Private right-of-way
pvmt.	Pavement
R. or Rge.	Range
Rad.	Radial or radius
R. C. P.	Reinforced concrete pipe
Rec.	Record
Rd.	Road
Rdwy.	Roadway
Ref.	Reference
Ret. Wall	Retaining wall
R. & O.	Rock and oil
R. P.	Reference point
R. R.	Railroad
R. R. R/W	Railroad right-of-way
Rt.	Right
R/W	Right-of-way
Ry.	Railway
S.	South or slope
San.	Sanitary
S. B. M. D.	Standard bench mark disk
S. D.	Sewer district or storm drain (use storm drain in preference to storm sewer)
Sdg.	Sounding
S/O	South of
Sec.	Section
Sec.	Seconds (when preceded by a figure)
S. E.	Southeast or semi-elliptical
S. W.	Southwest
S. & T.	Spike and tin
S. & W.	Spike and washer

Spec.	Specifications
Sp. M. H.	Special manhole
Sq.	Square
Sq. Ft.	Square foot or feet
S. S.	Sanitary sewer
Sta.	Station
⊡	Stadia station (sign)
Std.	Standard
Stk.	Stake
Stks.	Stakes
St.	Street
S. T.	Semitangent
Str. Gr.	Straight grade
Surv.	Survey
Struct.	Structure or structural
T. or Tan.	Tangent
Tel.	Telephone (pole)
Temp.	Temporary or temperature
Term.	Terminus
Terr.	Terrace
T. H.	Test hole
Tk.	Tack
T. L.	Total length of local depression traverse line
♉	Throw over (sign)
Topog.	Topographical
T. P.	Turning point
Tr.	Tract
Twp.	Township
△	Triangulation station (sign)
Trans.	Transition
T.S.	Traffic signal
U. S. C. E.	United States Corps of Engineers
U. S. C. & G. S.	United States Coast & Geodesic Survey
U. S. G. S.	United States Geological Survey (datum)
V. C.	Vertical curve
Vert.	Vertical
W.	West
W. C.	Witness corner

Wk.	Walk
W. L.	Water line
W/O	West of
W.P.	Witness post
W. S.	Wearing surface
X-sect.	Cross section
Yd.	Yard

Chapter 19

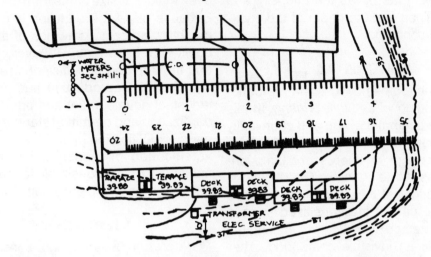

Reproduction of Drawings

PLENTY OF WORK IS REQUIRED TO PREPARE even the simplest drawings. The original drawings of any project is often very valuable and must be guarded against wear or becoming lost. One simple solution to this problem is to make copies of the original drawings for distribution to those who need them for reference.

Take this book for example. Many hours of work went into the research, writing, editing, typesetting, layout, proofreading, and printing. The original manuscript represents many thousands of dollars worth of time, money, and effort. Because thousands of copies have been printed from the original, each copy can be sold for an affordable sum. The reproduction of drawings is handled in a similar manner.

BLUEPRINTING

Blueprinting has been the accepted method of reproducing drawings from tracing paper for many years. Basically, the process of making a blueprint consists of placing a special paper coated with chemicals sensitive to sunlight in a glass frame with its coated side up. The original drawing on tracing paper is laid over the coated paper, and both are held together by the glass frame.

The entire assembly—paper, frame, and drawing—is then subjected to sunlight for a few minutes. The coated paper is thoroughly washed in clean water and hung up to dry. If the exposure has been timed correctly, the coated surface of the paper is now a clear, dark blue color except where it was covered by the lines on the drawing. That section is white.

An electric blueprinting machine is also used for making blueprints. The coated paper along with the original drawing passes around a glass cylinder containing electric lamps. The speed at which the paper travels can be adjusted as needed to suit the quality of the drawing, as far as the intensity and size of the lines and the depth of the background are concerned.

WHITEPRINTING

Most drafting rooms now use the reproduction method known as *whiteprinting*. In this method the original drawing and chemically treated paper are exposed to a high-intensity discharge or fluorescent lamp enclosed in a suitable housing. The ultraviolet light from either of these two light sources reduces the part of the sensitized surface that is unprotected by the lines of the original drawing into an invisible compound. After exposure, the print (or chemically treated paper) is then subjected to ammonia vapors that develop the sensitized lines. The finished print is true to scale and ready for immediate use. Depending on the type of paper used, the printed lines may be red, brown, blue, or black on a white to light/blue background.

MICROFILM

The microfilm process has recently become popular for reproducing and preserving drawings. Because of the small size of the film, much space is saved. Shipping charges can be greatly reduced.

Convert the original drawing to a microfilm frame by a special camera mounted on a frame over a platform such as a table. The camera reduces a drawing so that it will fit on the microfilm; the reduction can be varied by changing the height of the camera above the drawing.

After the film has been exposed, a processor develops it. It is then mounted on some type of card or frame. Once the film has been mounted, the image can be blown up by a special viewer. Some viewers will even make full-size prints from the microfilm. Because plenty of reduction is used in the microfilm process, drawings of the highest quality are important, with every line sharp and the lettering of sufficient size.

Graphic reproduction has accelerated industrial development and permitted standardization and permitted standardization of parts, simplification of effort, and greater diversification of industry. It is no longer necessary or even desirable for each industrial unit to be completely self-sufficient.

PHOTOGRAPHY

Photography uses the light-sensitive properties of the silver and silver nitrate compounds. The sensitized solution can be coated on practically any base material. The process lends itself easily to reduction and enlargements of graphic illustrations. It is used basically for the development of permanent reproductions, the rejuvenation of old graphic illustrations, and the development of translucent master copies from opaque originals.

Some sensitized coatings have to be handled under darkroom conditions, but others can be handled in subdued room light. Some sensitized materials are exposed when in direct contact with the graphic subject. Others have this subject projected on them from a distance. Certain sensitized materials create a positive print from a positive original. Others result in a negative print from a positive original. Several sensitized materials are exposed by having the light pass through the original to the sensitized coating. Others are exposed as the light passes through the sensitized coating first and then bounces back from the original to the sensitized coating. Several sensitized coatings require the light to pass through a yellow filter; others do not.

In every case the development after exposure is by standard photographic materials, water, hypo, and fixer. The reproductions must be dried before use.

Just as some chemicals show a marked change when subjected to light radiation, others are affected by heat radiation. In this process the base material is coated with a heat-sensitive chemical. When a graphic illustration is held between a low-grade heat source and the coating, the dark areas (lines) permit the sensitized coating to absorb more heat than the blank areas. The excess heat absorbed changes the color of the chemical coating to a blue-black, and the resultant reproduction is a positive copy of the original.

The disadvantages of this process at present are limitation in size and critical heat range. Also, some nonmetallic colors cannot be reproduced.

MAKING ALTERATIONS ON DRAWINGS

Drawings that require alteration are sometimes reproduced directly from the unchanged original on an intermediate. This intermediate is usually in the form of a *sepia* reproduction on which unwanted lines may be removed by the application of correction fluid. New data is then drawn in the correct area. The corrected drawing may then be reproduced by whiteprinting to deliver any number of copies.

Sometimes an original drawing with errors may be edited and then cut and pasted to make a new reproducible drawing. This saves much time and labor.

Another method of correcting original drawings for reproducing is to block certain portions of the original drawing. The remaining portion can be reproduced as usual with the blocked-out area invisible on the print. An example of this use could be for a building design for one of the chain businesses. All the buildings for a certain establishment, such as Pizza Hut, have a very similar floor plan and layout. The lots on which these buildings are constructed may be shaped so that the location of entry doors may have to be changed slightly. Other than this minor change (and the plot plan), the remaining structure will stay the same. The original drawing will be printed on a sepia sheet, which acts as an intermediate. This sepia print can then be blocked out, eradicated, or otherwise edited out. The new additions are then drawn on the sepia, and any number of prints may be produced from the new sepia. The original will be kept on file to be used with other building projects, where other changes may be made in a similar manner.

PHOTOCOPYING

Photocopying works similar to conventional photography in that a picture of the original drawing is taken. The machine (photocopier) then produces as many prints as desired. This type of copier makes it possible to reduce the original to any size. The process is also quicker than most other methods of reproducing working drawings.

Photocopying has many other advantages that can save much time in the drafting room. Sometimes original drawings become creased, stained, or worn to the point that they cannot be satisfactorily reproduced, revised, or microfilmed. The drawing usually must be redrawn by qualified draftsmen. They are photographic materials and techniques that can be used to restore old worn drawings. Figure 19-1 shows an example of a worn drawing, and Fig. 19-2 shows this same drawing after it has been restored by photocopying techniques.

During the design stage of any project, considerable revisions take place. For smaller projects where only one or two small drawings are necessary, it is usually simple to merely redraw the sheets and add or delete the changes. For the larger, more complex projects—sometimes requiring dozens of drawings sheets—such redrawing can be very costly. Again, the photocopier can save much valuable time. Rather than retrace a drawing that needs only a few revisions, a photostat print can be made. The unwanted part can be cut out, blocked out, etc. After taping the remaining drawing to a new drawing form, the composite is photographed and printed on a special base film. The revisions can then be made on this film. Much time is saved by redrawing only these portions actually needing revision.

When it becomes necessary to change the size of an original drawing to save file space or cut printing and postage costs, a photocopier can do the job at the flick of a switch. These machines can also enlarge drawings to bring out certain minute details for greater legibility or to make certain revisions easier. This size-changing feature of photocopiers has other advantages, one of

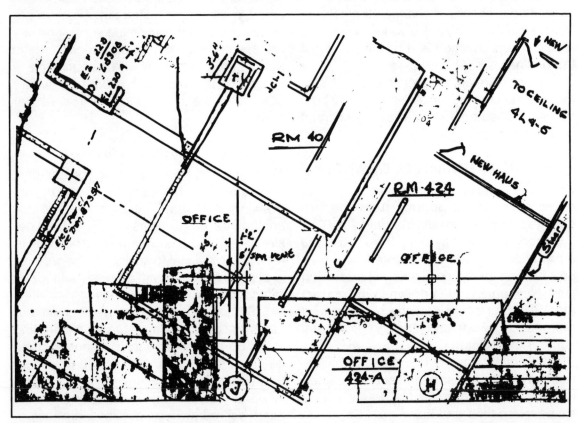

Fig. 19-1. A drawing of this sort is not suitable for reproduction by conventional means. It will have to be redrawn or else restored (courtesy Kodak).

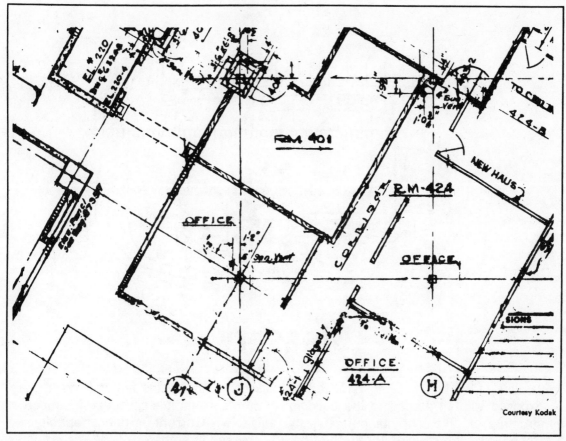

Fig. 19-2. This is the same drawing that appeared in Fig. 19-1 after being restored by photographic techniques (courtesy Kodak).

which is saving time in making detail drawings. Some drafting rooms are now making freehand sketches of certain items in extra-large size. The sketches look rather rough in this size, depending on the talent of the person doing the drawing. When the sketches, are reduced several times (say, one fourth regular size), the lines become straighter (to the eye). The sketches are usually suitable for publication. In most cases sketches of this type can save at least half the usual time required when straightedges and templates must be used.

The main disadvantage of photocopies is the relatively high cost, and the drafting or engineering department must have sufficient work to make the renting or purchasing of these machines worthwhile. If there is only an occasional use for a photocopier, there are businesses that have them available for small jobs.

If you want to make only a few copies of a particular project, a conventional camera may suffice. Merely use a close-up lens on your camera, snap the photo, and then have a few 8 × 10 prints made from your negative. Special printing papers are available for this purpose.

265

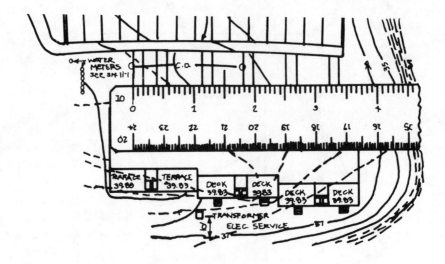

Glossary

accessible—Capable of being removed or exposed without damaging the building structure or finish, or not permanently closed in by the structure or finish of the building.

aggregate—Inert material mixed with cement and water to produce concrete.

air cleaner—Device used for removal of airborne impurities.

air diffuser—Air distribution outlet designed to direct airflow into desired patterns.

air entrained concrete—Concrete in which a small amount of air is trapped by addition of a special material to produce greater durability.

ambient temperature—Temperature of fluid (usually air) that surrounds an object on all sides.

American bond—Brickwork pattern consisting of five courses of stretchers followed by one bonding course of headers.

ammeter—An electric meter used to measure current. It is calibrated in amperes.

ampacity—Current-carrying capacity expressed in amperes.

amplification—Procedure of expanding the strength of a signal.

amplifier—An instrument used to increase the signal voltage, current, or power.

anode—Positive electrode or terminal; the plate of a vacuum tube. In some cases it is the most positive electrode.

antisiphon trap—Trap in a drainage system designed to preserve a water seal by defeating siphonage.

appliances—Utilization equipment normally built in standardized sizes or types and installed or connected as a unit to perform one or more functions such as clothes washing, air conditioning, etc.

apron—Piece of horizontal wood trim under the sill of the interior casing of a window.

areaway—Open space below the ground level immediately outside a building. It is enclosed by substantial walls.

arrester—Wire screen secured to the top of an incinerator to confine sparks and other products of burning.

ashlar—Squared and dressed stones used for facing a masonry wall; short upright wood pieces extending from the attic floor to the rafters forming a dwarf wall.

attachment plug (plug cap) (cap)—A device that, on insertion in a receptacle, establishes a connection between the conductors of the attached flexible cord and the conductors connected permanently to the receptacle.

attenuation—The decrease in the strength of a signal.

automatic—Self-acting, operating by its own mechanism when actuated by some impersonal influence, such as a change in current strength, pressure, temperature, or mechanical configuration.

backfill—Loose earth placed outside foundation walls for filling and grading.

back pressure—Pressure in the low side of a refrigerating system; also called suction pressure or low side pressure.

balloon framing—System of small house framing; two by fours extending two stories with inch by quarter ledger strips notched into the studs to support the second-story floor beams.

balustrade—Protective or decorative railing.

bargeboard—Ornamented board covering the roof boards and projecting over the slope of the roof.

barometer—Instrument for measuring atmospheric pressure.

base—One of the regions or terminals of a transistor.

batten—Narrow wood strip used to cover joints; also, a wood strip used to secure adjoining boards.

batter—Slope of the exposed face of a retaining wall.

bead—Narrow projecting molding with a rounded surface; or in plastering, a metal strip embedded in plaster at the projecting corner of a wall.

beam—A horizontal member of wood, reinforced concrete, steel, or other material used to span the space between posts, columns, girders, or over an opening in a wall.

bearing plate—Steel plate placed under one end of a beam or truss for load distribution.

bearing wall—Wall supporting a load other than its own weight.

bed—Place or material in which stone or brick is laid; horizontal surface of positioned stone; lower surface of brick, stone, or tile.

belt course—Decorative horizontal band of masonry.

bench mark—Point of reference from which measurements are made.

bias vacuum tube—Difference of potential between control grid and cathode. Transistor—difference of potential between base and emitter and base and collector. Magnetic amplifier—level of flux density in the magnetic amplifier core under no-signal condition.

billet—Heavy steel plate placed on concrete providing support for a column.

bimetal strip—Temperature regulating or indicating device that works on the principle that two dissimilar metals with unequal expansion rates, welded together, will bend as temperature changes.

block bridging—Solid wood members nailed between joists to stiffen a floor.

boiler—Closed container in which a liquid may be heated and vaporized.

boiling point—Temperature at which a liquid boils or generates bubbles of vapor when heated.

bonding jumper—A reliable conductor used to ensure the electrical conductivity between metal parts required to be electrically connected.

bore—A hole machined in a piece that is made with a boring tool.

boss—A raised ornamentation; an ornamental projecting block used in architecture.

braced frame—System of wood house framing using posts, girts, and braces.

branch circuit—That portion of a wiring system extending beyond the final overcurrent device protecting a circuit.

bridging—System of bracing between floor beams to distribute floor load.

British thermal unit (Btu)—Quantity of heat required to raise the temperature of 1 pound of water 1 degree Fahrenheit.

buck—Rough wood door frame placed on a wall or partition to which the door moldings are attached; completely fabricated steel door frame set in a wall or partition to receive the door.

burner—Device in which combustion of fuel takes place.

bus bar—The heavy copper or aluminum bar used on switchboards to carry current.

butane—Liquid hydrocarbon commonly used as fuel for heating purposes.

buttress—Projecting structure built against a wall to give it greater strength.

bypass—Passage at one side of or around a regular passage.

caisson—Sunken panel in a ceiling, contributing to a pattern.

calorie—Heat required to raise temperature of 1 gram of water 1 degree centigrade.

cantilever—Projecting beam or member supported at only one end.

cant strip—Beveled strip placed in the angle between the roof and an abutting wall to avoid a sharp bend in the roofing material; strip placed under the lowest row of tiles on a roof to give it the same slope as the rows above it.

capacitor—An apparatus consisting of two conducting surfaces separated by an insulating material. It stores energy, blocks the flow of direct current, and permits the flow of alternating current to a degree depending on the capacitance and frequency.

carbon dioxide—Compound of carbon and oxygen that is sometimes used as a refrigerant.

carbon filter—Air filter using activated carbon as air-cleansing agent.

casement window—Window sash opening on hinges secured to the side of the window opening.

cathode—The electron-emitting electrode of an electron tube or semiconductor device.

cavity wall—Wall built of solid masonry units arranged to provide airspace within the wall.

Celsius—German word for centigrade; the metric system's temperature scale.

centering—Temporary wood framing supporting concrete forms.

centigrade scale—Temperature scale used in metric system. Freezing point of water is 0 degrees; boiling point is 100 degrees.

chamfer—Bevel edge surface area produced by cutting away the external angle formed by two faces of stone or lumber.

chase—Recess in inner face of masonry wall providing space for pipes and/or ducts.

check valve—A device that permits fluid flow only in one direction.

chimney effect—Tendency of air or gas to rise when heated.

choke coil—A coil used to limit the flow of alternating current while permitting direct current to pass.

circuit—A closed path through which current flows from a generator, through various components, and back to the generator.

circuit breaker—A resettable fuse-like device designed to protect a circuit against overloading.

coaxial cable—A cable consisting of two conductors concentric with and insulated from each other.

code installation—An installation that conforms to the local code and/or the national code for safe and efficient installation.

cold cathode—A cathode that does not depend on heat for electron emission.

collar beam—Horizontal tie beam connecting rafters considerably above the plate.

collector—The part of a transistor that collects electrons.

column—Vertical load-carrying member of a structural frame.

comfort zone—Area on psychrometric chart that shows conditions of temperature, humidity, and sometimes air movement in which most people are comfortable.

commutator—Device used on electric motors or generators to maintain a unidirectional current.

compressor—The pump of a refrigerating mechanism that draws a vacuum or low pressure on the cooling side of a refrigerant cycle and squeezes or compresses the gas into the high pressure or condensing side of the cycle.

concealed—Rendered inaccessible by the structure or finish of the building. Wires in concealed raceways are considered concealed, even though they may become accessible by withdrawing them.

conductance—The ability of material to carry an electric current.

conductor, bare—Having no covering or insulation whatsoever.

conductor, covered—A conductor having one or more layers of nonconducting materials that are not recognized as insulation under the National Electric Code.

conductor, insulated—A conductor covered with material recognized as insulation.

connector, pressure (solderless)—A connector that establishes the connection between two or more conductors or between one or more conductors and a terminal by mechanical pressure and without the use of solder.

continuous load—A load in which the maximum current is expected to continue for three hours or more.

contour line—On a land map denoting elevations, a line connecting points with the same elevation.

control—Automatic or manual device used to stop, start, and/or regulate flow of gas, liquid, and/or electricity.

control, temperature—A thermostatic device that automatically stops and starts a motor, the operation of which is based on temperature changes.

controller—A device or group of devices that serves to govern in some predetermined manner the electric power delivered to the apparatus to which it is connected.

convection—Transfer of heat by movement or flow of a fluid or gas.

cooling tower—Device that cools water by water evaporation in air. Water is cooled to the wet bulb temperature of air.

coping—The highest course of a masonry wall bedded on the parapet wall.

corbeling—Projecting courses of brick stepped out from the face of the wall forming a bracket for the wall above.

core—The portion of a foundry mold that shapes the interior of a hollow casting.

cored hole—At the time of casting, a sand core is placed in the mold so that the metal flows around it. When the casting is cold, the sand core is broken away, leaving the hole in the casting.

cornice—The part of a roof that projects from the outside wall. It is used to carry the gutter and serves as protection for the wall.

counterbore—A tool that enlarges an already-machined round hole to a certain depth. The pilot of the tool fits in the smaller hole, and the larger part counterbores or makes the end of the hole larger.

counterflashing—Sheet metal strip in the form of an inverted L built into a wall to overlap the flashing and make the roof watertight.

cramp—Iron rod with ends bent to a right angle; used to hold blocks of stone together.

crawl space—Shallow space between the first tier of beams and the ground (no basement).

cricket—Small false roof to throw off water from behind an obstacle.

critical temperature—Temperature at which vapor and liquid have the same properties.

cryogen—A substance that becomes a superconductor at extremely cold temperatures.

current—The rate of transfer of electricity.

curtain wall—Nonbearing wall built between piers or columns for the enclosure of the structure; not supported at each story.

cycle—An interval of time during which a sequence of a recurring succession of events is completed.

damper—Valve for controlling air flow.

dead man—Reinforced concrete anchor set in earth and tied to the retaining wall for stability.

deflecting plate—The part of a certain type of electron tube that provides an electrical field to produce deflection of an electron beam.

deflection—Deviation of the central axis of a beam from normal when the beam is loaded.

demand factor—In any system or part of a system, the ratio of the maximum demand of the system, or part of the system, to the total connected load of the system, or part of the system under consideration.

density—Closeness of texture or consistency.

dentil—One of a series of small projecting rectangular blocks under a cornice.

detection—The process of obtaining the separation of the modulation, component from the received signal.

dielectric—An insulator or a term referring to the insulation between the plates of a capacitor.

diode—A device having two electrodes, the cathode and the plate or anode—used as a rectifier and detector.

disconnecting means—A device, a group of devices, or other means whereby the conductors of a circuit can be disconnected from their supply source.

dormer window—Extension from a sloped roof with a vertical window.

double-hung window—Window consisting of two sashes sliding vertically in adjoining grooves.

double-strength glass—One-eighth inch thick sheet glass (single strength glass is 1/10 inch thick).

draft gauge—Instrument used to measure air movement.

draft indicator—An instrument used to indicate or measure chimney draft or combustion gas movement.

drill—A circular tool used for machining a hole.

drip—Projecting horizontal course sloped outward to throw water away from a building.

dry bulb—An instrument with a sensitive element that measures ambient (moving) air temperature.

drywall—Interior wall construction consisting of plasterboard, wood paneling, or plywood nailed directly to the studs without application of plaster.

duct—A tube or channel through which air is conveyed or moved.

duty, continuous—A service requirement that demands operation at a substantially constant load for an indefinitely long time.

duty, intermittent—A service requirement that demands operation for alternate intervals of load and no load, load and rest, or load, no load, and rest.

duty, periodic—A type of intermittent duty in which the load conditions regularly recur.

duty, short-time—A requirement of service that demands operations at loads and for intervals of time, both of which may be subject to wide variation.

dwarf partition—Partition that ends short of the ceiling.

dynamometer—Device for measuring power output or power input of a mechanism.

effective temperature—Overall effect on a person of air temperature, humidity, and air movement.

electric defrosting—Use of electric resistance heating coils to melt ice and frost off evaporators during defrosting.

electric heating—House heating system in which heat from electrical resistance units is used to heat rooms.

electric water valve—Solenoid type (electrically operated) valve used to turn water flow on and off.

electrode—A conducting element used to emit, collect, or control electrons and ions.

electrolytic condenser-capacitor—Plate or surface capable of storing small electrical charges. Common electrolytic condensers are formed by rolling thin sheets of foil between insulating materials. Condenser capacity is expressed in microfarads.

electromotive force (emf) voltage—Electrical force that causes current (free electrons) to flow or move in an electrical circuit. Unit of measurement is the volt.

electron—The subatomic particle that carries the unit negative charge of electricity.

electron emission—The release of electrons

271

from the surface of a material into surrounding space due to heat, light, high voltage, or other causes.

elevation—Drawing showing the projection of a building on a vertical plane.

emitter—The part of a transistor that emits electrons.

enclosed—Surrounded by a case that will prevent anyone from accidentally touching live parts.

end bearing pile—Pilt acting like a column; the point has a solid bearing in rock or other dense material.

epitaxial—A very significant thin film type of deposit for making certain devices in microcircuits involving a realignment of molecules.

equipment—A general term including material, fittings, devices, appliances, fixtures, apparatus, and the like used as part of, or in connection with, an electrical installation.

evaporation—A term applied to the changing of a liquid to a gas. Heat is absorbed in this process.

evaporator—Part of a refrigerating mechanism in which the refrigerant vaporizes and absorbs heat.

expansion bolt—Bolt with a casing arranged to wedge the bolt into a masonry wall to provide an anchor.

expansion joint—Joint between two adjoining concrete members arranged to permit expansion and contraction with temperature changes.

expansion valve—A device in a refrigerating system that maintains a pressure difference between the high side and low side and is operated by pressure.

exposed (as applied to live parts)—Live parts that a person could inadvertently touch or approach nearer than a safe distance. This term is applied to parts not suitably guarded, isolated, insulated.

exposed (as applied to wiring method)—Not concealed; externally operable; capable of being operated without exposing the operator to contact with live parts.

eyelet—Something used on printed circuit boards to make reliable connections from one side of the board to the other.

facade—Main front of a building.

face—An operation that machines the sides or ends of the piece.

face brick—Brick selected for appearance in an exposed wall.

face of a gear—That portion of the tooth curve above the pitch circle and measured across the rim of the gear from one end of the tooth to the other.

factor of safety—Radio of ultimate strength of material to maximum permissible stress in use.

fail-safe control—Device that opens a circuit when the sensing element fails to operate.

fan—A radial or axial flow device used for moving or producing artificial currents of air.

F.A.O.—This symbol on a mechanical drawing means that the piece is machined or finished all over.

farad—The basic unit of capacitance.

feedback—The process of transferring energy from the output circuit, of a device back to its input.

feeder—The conductors between the service equipment, or the generator switchboard of an isolated plant, and the branch circuit overcurrent device.

filament—A cathode in the form of a metal wire in an electron tube.

fillet—The rounded corner or portion that joins two surfaces which are at an angle to each other.

filter—A porous article through which a gas or liquid is passed to separate out matter in suspension; a circuit or devices that pass one frequency or frequency band while blocking others, or vice versa.

finish plaster—Final or white coat of plaster.

firebrick—Brick made to withstand high temperatures that is used for lining chimneys, incinerators, and similar structures.

fireproof wood—Chemically treated wood; fire-resistive, used where incombustible materials are required.

fire-rated doors—Doors designed to resist standard fire tests and labeled for identification.

fire-stop—Incombustible filler material used to block interior draft spaces.

fitting—An accessory such as a locknut, bushing, or other part of a wiring system that is intended primarily to perform a mechanical rather than an electrical function.

flapper valve—The type of valve used in refrigeration compressors that allows gaseous refrigerants to flow in only one direction.

flare—Copper tubing is often connected to parts of a refrigerating system by flared fittings. They require that the end of the tube be expanded at about a 45-degree angle.

flashing—Strips of sheet metal bent into an angle between the roof and wall to make a watertight joint.

flash point—Lowest temperature at which vapors above a volatile combustible substance ignite in air when exposed to flame.

flat slab construction—Reinforced concrete floor construction of uniform thickness; eliminates the drops of beams and girders.

Flemish bond—Pattern of bonding in brickwork consisting of alternate headers and stretchers in the same course.

flitch beam—Built-up beam consisting of a steel plate sandwiched between wood members and bolted.

float valve—Type of valve that is operated by a sphere or pan which floats on a liquid surface and controls the level of liquid.

flooding—Act of filling a space with a liquid.

flow meter—Instrument used to measure velocity or volume of fluid movement.

footing—Structural unit used to distribute loads to the bearing materials.

forced convection—Movement of fluid by mechanical force such as fans or pumps.

foundation—Composed of footings, piers, foundation walls (basement walls), and any special underground construction necessary to properly support the structure.

frequency—The number of complete cycles an alternating electric current, sound wave, or vibrating object undergoes per second.

friction pile—Pile with supporting capacity produced by friction with the soil in contact with the pile.

frost line—Deepest level below grade to which frost penetrates in a geographic area.

furring—Thin wood, brick, or metal applied to joists, studs, or wall to form a level surface (as for attaching wallboard) or airspace.

fuse—Electrical safety device consisting of a strip of fusible metal in a circuit that melts when the current is overloaded.

fusible plug—A plug or fitting made with a metal of a known low melting temperature; used as a safety device to release pressures in case of fire.

gain—The ratio of output to input power, voltage, or current, respectively.

gambrel roof—Roof with its slope broken by an obtuse angle.

garage—A building or portion of a building in which one or more self-propelled vehicles carrying volatile, flammable liquid for fuel or power are kept.

garden bond—Bond formed by inserting headers at wide intervals.

gas—Vapor phase or state of a substance.

gate—A device that makes an electronic circuit operable for a short time.

girder—A large beam made of wood, steel, or reinforced concrete.

girt—Heavy timber framed into corner posts as support for the building.

government anchor—A V-shaped anchor usually made of ½-inch round bars to secure the steel beam to masonry.

grade beam—Horizontal, reinforced concrete beam between two supporting piers at or below ground supporting a wall or structure.

grid—An electrode having one or more openings for the passage of electrons or ions.

grid leak—A resistor of high ohmic value connected between the control grid and the cathode in a grid-leak capacitor detector circuit and used for automatic biasing.

grillage—Steel framework in a foundation designed to spread a concentrated load over a wider area; generally enclosed in concrete.

grille—An ornamental or louvered opening placed at the end of an air passageway.

groined ceiling—Arched ceiling consisting of two intersecting curved, planes.

grommet—A plastic metal or rubber doughnut-shaped protector for wires or tubing as they pass through a hole in an object.

ground—A large conducting body (as the earth) used as a common return for an electric circuit and as an arbitrary zero of potential.

ground coil—A heat exchanger buried in the ground that may be used either as an evaporator or a condenser.

grounded conductor—A system or circuit conductor that is intentionally grounded.

grounding conductor—A conductor used to connect equipment or the grounded circuit of a wiring system to a grounding electrode.

grounds—Narrow strips of wood nailed to walls as guides to plastering and as a nailing base for interior trim.

grout—Thin mortar that is fluid enough to be poured into narrow spaces.

gusset—A plate or bracket for strengthening an angle in framework.

half-lap joint—Joint formed by cutting away half the thickness of each piece.

H beam—Steel beam with wider flanges than an I beam.

header—Brick laid with an end exposed in the wall; wood beam set between two trimmers and carrying the tail beams.

heading bond—Pattern of brick bonding formed with headers.

head pressure—Pressure that exists in the condensing side of a refrigerating system.

heat—Added energy that causes substances to rise in temperature; energy associated with random motion of molecules.

heat exchanger—Device used to transfer heat from a warm or hot surface to a cold or cooler surface. Evaporators and condensers are heat exchangers.

heating value—Amount of heat that may be obtained by burning a fuel; usually expressed in Btu per pound or gallon.

heat load—Amount of heat, measured in Btu, that is removed during a period of 24 hours.

heat of compression—Mechanical energy of pressure transformed into heat energy.

heat pressure control—Pressure-operated control that opens an electrical circuit if high side pressure becomes excessive.

heat pump—A compression cycle system used to supply heat to a temperature—controlled space, which can also remove heat from the same space.

heat transfer—Movement of heat from one body or substance to another. Heat may be transferred by radiation, conduction, convection, or a combination of these.

henry—The basic unit of inductance.

hermetic motor—Compressor drive motor sealed within the same casing that contains the compressor.

hermetic system—Refrigeration system that has a compressor driven by a motor contained in a compressor dome or housing.

high pressure cutout—Electrical control switch operated by the high side pressure that automatically opens an electrical circuit if too high head pressure or condensing pressure is reached.

high side—Parts of a refrigerating system that are under condensing or high side pressure.

high tension bolts—High strength steel bolts tightened with calibrated wrenches to high tension; used as a substitute for conventional rivets in steel frame structures.

hip jack rafter—Short rafter extending from the plate to the hip ridge.

hip rafter—Rafter extending from the plate to the ridge forming the angle, to a hip roof.

hip roof—Roof with sloping sides and sloping end.

hot gas bypass—Piping system in a refrigerating unit that moves hot refrigerant gas from a condenser into the low pressure side.

hot junction—That part of the thermoelectric circuit which releases heat.

hot wire—A resistance wire in an electrical relay that expands when heated and contracts when cooled.

humidity—Moisture, dampness. Relative humidity is the ratio of the quantity of vapor present in the air to the greatest amount possible at a given temperature.

hydrometer—Floating instrument used to measure specific gravity of a liquid. Specific gravity is the ratio of the density of a material to the density of a substance accepted as a standard.

hydronic—Type of heating system that circulates a heated fluid, usually water, through baseboard coils. Circulating pump is usually controlled by a thermostat.

hygrometer—An instrument used to measure the degree of moisture in the atmosphere.

I beam—Rolled steel beam or built-up beam of an I section.

ignition transformer—A transformer designed to provide a high voltage current.

impedance—The total opposition offered to the flow of an alternating current.

incombustible material—Material that will not ignite or actively support combustion in a surrounding temperature of 1200 degrees Fahrenheit during an exposure of five minutes; also, material that will not melt when the temperature of the material is maintained at 900 degrees Fahrenheit for at least five minutes.

inductance—The property of a circuit or two neighboring circuits that determines how much voltage will be induced in one circuit by a change of current in either circuit.

inductor—A coil.

infrared lamp—An electrical device that emits infrared rays—invisible rays just beyond red in the visible spectrum.

insulation, thermal—Substance used to retard or slow the flow of heat through a wall or partition.

integrated circuit—A circuit in which different types of devices such as resistors, capacitors, and transistors are made from a single piece of material and then connected to form a circuit.

IR drop—The voltage drop across a resistance due to the flow of current through the resistor.

isolated—Not readily accessible to persons unless special means for access are used.

isothermal—Changes of volume or pressure under conditions of constant temperature.

jack rafter—Short rafter used in hip or valley framing.

jamb—Upright member forming the side of a door or window opening.

joists and plank—Pieces of lumber with nominal dimensions of 2 to 4 inches in thickness by 4 inches and wider, of a rectangular cross section and graded with respect to their strength in bending when loaded either on the narrow face as a joist or on the wide face as a plank.

junction box—Group of electrical terminals housed in a protective box or container.

junior beam—Light weight structural steel sections rolled to a full I beam shape.

keyway—Groove formed in the top of a footing to anchor the foundation wall above; any groove formed in poured concrete to receive a succeeding pour.

kilometer—A metric unit of linear measurement equal to 1000 meters.

kilowatt—Unit of electrical power equal to 1000 watts.

king post—Central vertical tie in a truss.

knee brace—Diagonal member placed across the inside angle of framework to stiffen the frame.

lacquer—A protective coating or finish that dries to form a film by evaporation of a volatile constituent.

Lally column—Concrete-filled cylindrical steel structural column.

laminated wood—Wood built up of plies or laminations that have been joined either with glue or with mechanical fasteners. The plies usually are too thick to be classified as veneer, and the grain of all plies is parallel.

lap joint—Joint between two wood members in which the same width and depth of the members are retained.

latent heat—Heat given off or absorbed in a process (as vaporization or fusion) other than a change in temperature.

leader—Vertical sheet metal pipe conducting rain from the roof gutter.

ledger—In balloon framing, the board

notched into the exterior studs supporting the second-story floor beams.

let-in braces—In wood house framing, the diagonal braces notched into studs.

lighting outlet—An outlet intended for the direct connection of a lamp holder, lighting fixture, or pendant cord terminating in a lamp holder.

limit control—Control used to open or close electrical circuits as temperature or pressure limits are reached.

lintel—Horizontal steel member spanning an opening to support the load above.

liquid absorbent—A chemical in liquid form that has the property to absorb moisture.

liquid line—The tube that carries liquid refrigerant from the condenser or liquid receiver to the refrigerant control mechanism.

liquid receiver—Cylinder connected to a condenser outlet for storage of liquid refrigerant in a system.

liter—Metric unit of volume that equals 61.03 cubic inches.

live load—Any load on a structure other than a dead load includes the weight of persons occupying the building and free-standing material.

location, damp—A location subject to a moderate amount of moisture such as some basements, barns, cold-storage warehouses, and the like.

location, dry—A location not normally subject to dampness or wetness; a location classified as dry may be temporarily subject to dampness or wetness, as in the case of a building under construction.

location, wet—A location subject to saturation with water or other liquids.

lock seam—Joining of two sheets of metal consisting of a folded, pressed, and soldered joint.

loft—Upper floor of a business building; wide floor area without partitions.

logic—The arrangement of circuitry designed to accomplish certain objectives.

magnetic field—Space in which magnetic lines of force exist.

manometer—Instrument for measuring pressure of gases and vapors.

mansard roof—Roof having two slopes on all sides with the lower slope steeper than the upper one.

mat foundation—Continuous reinforced concrete foundation constructed under the entire building as a unit.

megohmmeter—An instrument for measuring extremely high resistances.

membrane waterproofing—System of waterproofing masonry walls with layers of felt, canvas, or burlap and pitch.

meter—Metric unit of linear measurement equal to 39.37 inches.

micrometer—A precision measuring instrument.

millwork—Finished wood products manufactured in planing mills such as window frames, windows, doors, stairways, interior trim, etc; does not include flooring, siding, or ceiling.

modulation— Varying the amplitude, frequency, or the phrase of a carrier wave.

moisture determination—The use of instruments and calculators to measure the relative or absolute moisture in an air conditioned space.

motor, capacitor—A single-phase induction motor with an auxiliary starting winding connected in series with a condenser for better starting characteristics.

motor control—Device to start and/or stop a

motor at certain temperature or pressure conditions.

mullion—Vertical member forming a division between adjoining windows.

multioutlet assembly—A type of surface or flush raceway designed to hold conductors and attachment plug receptacles.

muntin—A strip separating panes of glass in a sash.

nailing blocks—Wood members set on masonry to anchor other members to the masonry with nails or screws.

natural convection—Movement of a fluid or air caused by temperature differences.

neutron—An uncharged atomic particle having a mass approximately equal to that of the proton.

nonautomatic—Used to describe an action requiring personal intervention for its control.

nonbearing wall—Wall that carries no load other than its own weight.

noncode installation—A system installed where there are no local, state, or national codes in force.

normal charge—The thermal element charge that is part liquid and part gas under all operating conditions.

ohm—Unit of measurement of electrical resistance. One ohm exists when 1 volt causes a flow of 1 ampere.

ohmmeter—An instrument for measuring resistance in ohms.

Ohm's law—Mathematical relationship between voltage, current, and resistance in an electric circuit.

one-way slab—Concrete slab with reinforcing steel rods providing a bearing on two opposite sides only.

open circuit—An interrupted electrical circuit that stops the flow of electricity.

open web joist—Steel joists built up out of light steel shapes with an open latticed web.

orifice—Accurate size opening for controlling fluid flow.

oscillator—A device that produces an alternating or pulsating current or voltage electronically.

oscilloscope—A test instrument that shows visually on a screen the pattern representing variations in an electrical quantity.

outlet—In the wiring system, a point at which current is taken to supply utilization equipment.

outline lighting—An arrangement of incandescent lamps or gaseous tubes to outline and call attention to certain features, such as the shape of a building or the decoration of a window.

overload—Load greater than the load for which the system or mechanism was intended.

overload protection—The use of circuit breakers, relays, automatic limiters, and similar devices to protect equipment from overload damage by reducing current or voltage.

ozone—A triatomic form of oxygen that is formed naturally in the upper atmosphere by a photochemical reaction with solar ultraviolet radiation or generated commercially by a silent electric discharge in oxygen or air.

panelboard—A single panel or group of panel units designed for assembly in the form of a single panel; includes buses and may come with or without switches and/or automatic overcurrent protective devices for the control of light, heat, or power circuits of individual as well as aggregate capacity. It is designed to be placed in a cabinet or cutout box that is in or against a wall or partition and is accessible only from the front.

parapet—Part of a masonry wall extending above the roof.

pentode—An electron tube with five electrodes or elements.

Permalloy—An alloy of nickel and iron that is easily magnetized and demagnetized.

photoelectricity—A physical action wherein an electrical flow is generated by light waves.

piers—Masses of brick, stone masonry, or concrete used as supports.

pilaster—Flat square column attached to a wall and projecting about one-fifth of its width from the face of the wall.

piles—Long, slender members of wood, steel, or reinforced concrete driven into the ground to carry a vertical load.

pintle—Iron member used at the base of a wood post to anchor it into place; generally used with a cast-iron cap over the post below.

plate—The principal anode in an electron tube at which electrons collect.

plate girder—Built-up girder resembling an I beam with a web of steel plate and flanges of angle iron.

platform framing—System of wood frame house construction using wood studs one story high finished with a platform consisting of the underflooring for the next story.

plenum—Chamber or space forming a part of an air conditioning system.

plinth—Lowest member of a base or pedestal.

porcelain—Ceramic chinalike coating applied to steel surfaces.

post and girt—Wood framed buildings consisting of widely spaced load-bearing wood posts connected with horizontal members called girts.

potential—The difference in voltage between two points of a circuit. Frequently, one is assumed to be ground (zero potential).

potentiometer—An instrument for measuring an unknown voltage or potential difference by balancing it, wholly or in part, by a known potential difference produced by the flow of known currents in a network of circuits of known electrical constants.

power—The rate of doing work or expending energy.

power element—Sensitive element of a temperature-operated control.

power factor—Correction coefficient for ac power necessary because of changing current and voltage values.

precast concrete—Concrete units (such as piles or vaults) cast away from the construction site and set in place.

pressure—An energy impact on a unit area; force or thrust exerted on a surface.

pressure motor control—A device that opens and closes an electrical circuit as pressures change.

prestressed concrete—System for utilizing fully the compressive strength of concrete by bonding it with highly stressed tensile steel.

primary control—Device that directly controls operation of a heating system.

protector, circuit—An electrical device that will open an electrical circuit if excessive electrical conditions occur.

psi—Initials used to indicate pressure measured in pounds, per square inch.

pump down—The act of using a compressor or a pump to reduce the pressure in a container or a system.

purlin—Horizontal members of a roof that rest on roof trusses and support rafters.

pyrometer—Instrument for measuring high temperatures.

qualified person—One familiar with the construction and operation of the apparatus and the hazards involved.

queen post—One of two vertical posts in a roof truss.

quoin—Corner blocks of masonry; stone or brick set at the corner of a building in blocks forming a decorative pattern.

raceway—Any channel designed expressly for holding wire, cables, or bus bars and used solely for that purpose.

radiant heating—Heating system in which warm or hot surfaces are used to radiate heat into the space to be conditioned.

radiation—The process of emitting radiant energy in the form of waves or particles.

raked joint—Joint formed in brickwork by raking out some of the mortar an even distance from the face of the wall.

reamer—A finishing tool that finishes a circular hole more accurately than a drill.

receptacle (convenience outlet)—A contact device installed at an outlet for the connection of an attachment plug.

reciprocating—Action in which the motion is back and forth in a straight line.

rectifiers—Devices used to change alternating current to unidirectional current.

reflective insulation—Thin sheets of metal or foil on paper set in the exterior walls of a building to reflect radiant energy.

refrigerant—Substance used in refrigerating mechanism to absorb heat in an evaporator coil and to release heat in a condenser as the substance goes from a gaseous state back to a liquid state.

register—Combination grille and damper assembly covering on an air opening or end of an air duct.

relative humidity—Ratio of amount of water vapor present in air to greatest amount possible at same temperature.

relay—An electromechanical switching device that can be used as a remote control.

relief valve—Safety device to permit the escape of steam or hot water subjected to excessive pressures or temperatures.

remote-control circuit—Any electrical circuit that controls any other circuit through a relay or equivalent device.

resistance—The opposition that a device or material offers to the flow of current.

resistor—A circuit element whose chief characteristic is resistance.

resonance—In a circuit containing both inductance and capacitance, a condition in which the inductive reactance is equal to and cancels out the capacitance reactance.

return—The end railings of a fire escape balcony.

reveal—Space between window or door frame and the outside edge of the wall.

ridge pole—Highest horizontal member of a roof receiving upper ends of rafters.

riser—Upright member of a stair extending from tread to tread.

riser valve—Device used to manually control flow of refrigerant in vertical piping.

rotor—Rotating part of a mechanism.

roughing in—Installation of all concealed plumbing pipes; includes all plumbing work done before setting of fixtures or finishing.

rowlock—Pattern of brickwork consisting of a course of brick laid on edge with ends exposed.

running bond—Brick bond consisting entirely of stretchers.

rusticated work—Squared stones with edges beveled or grooved to make the joints stand out.

saddle—Short horizontal member set on top of a post to spread the load of the girder over it; piece of wood, stone, or metal placed under a door.

saddle valve—Valve body shaped so it may be silver brazed to a refrigerant tubing surface.

safety control—Device that will stop the refrigerating unit if unsafe pressures and/or temperatures are reached.

safety motor control—Electrical device used to open a circuit if the temperature, pressure, and/or the current flow exceed safe conditions.

safety plug—Device that will release the contents of a container above normal pressure conditions and before rupture pressures are reached.

sash balance—A device—dispensing with weights, pulleys, or cord—operated with a spring to counterbalance a double-hung window sash.

saturation—The condition existing in a circuit when an increase in the driving signal does not produce any further change in the resultant effect.

scavenger pump—Mechanism used to remove fluid from sump or containers.

scratch coat—First coat of plaster forming the key to the lath and the base for the second or brown coat.

sealed motor compressor—A mechanical compressor consisting of a compressor and a motor, both of which are enclosed in the same sealed housing, with no external shaft or shaft seals, and with the motor operating in the refrigerant atmosphere.

seat angle—Small steel angle riveted to one member to support the end of a beam or girder.

self-inductance—Magnetic field induced in the conductor carrying the current.

semiconductor—A material that has electrical properties of current flow between a conductor and an insulator.

sensible heat—Heat that causes a change in temperature of a substance.

sensor—A material or device that goes through a physical change or an electronic characteristic change as conditions change.

separator—Sections of steel pipe forming spacers between I beams bolted together serving as a structural unit.

sequence controls—Devices that act in series or in time order.

service—The conductors and equipment used for delivering energy from the electricity supply system to the wiring system of the premises served.

service cable—The service conductors made up in the form of a cable.

service conductors—The supply conductors that extend from the street main or transformers to the service equipment of the premises being supplied.

service drop—The overhead service conductors from the last pole, or other aerial support, to and including the splices, if any, that connect to the service-entrance conductors at the building or other structure.

service-entrance conductors, underground system—The service conductors between the terminals of the service equipment and the point of connection to the service lateral.

service equipment—The necessary equipment, usually consisting of a circuit breaker or switch and fuses and their accessories, located near the point of entrance of supply conductors to a building and intended to constitute the main control and cutoff means for the supply to that building.

service lateral—The underground service conductors between the street main, including any risers at a pole or other structure or from transformers, and the first point of connection to the service-entrance conductors in a terminal box, meter, or other enclosure with adequate space, inside or outside the building wall. Where there is no terminal box, meter, or other enclosure with adequate space, the point of connection is the entrance point of the service conductors into the building.

service raceway—The rigid metal conduit, electrical metallic tubing, or other raceway that encloses the service-entrance conductors.

service valve—A device to be attached to a system that provides an opening for gauges and/or charging lines.

setting (of circuit breaker)—The value of the current at which the circuit breaker is set to trip.

shaded pole motor—A small ac motor used for light start loads that has no brushes or commutator.

sheathing—First covering of boards or paneling nailed to the outside wall of a frame building or on a timber roof.

sheave beams—Steel beams forming the overhead supports for an elevator.

shim—Thin piece of material used to bring members to an even or level bearing.

shiplap—Wood boards cut with a rabbet or groove at opposite edges to provide an overlapping joint.

short circuit—An often unintended low-resistance path through which current flows around, rather than through, a component or circuit.

short cycling—Refrigerating system that starts and stops more frequently than it should.

shoved joint—Mortar joint produced by laying brick in a thick bed of mortar and forming a vertical joint by pushing the brick against the brick already laid in the same course.

shroud—Housing over a condenser or evaporator.

siding—Finishing material nailed to the sheathing of wood frame buildings and forming the exposed surface.

sight glass—Glass tube or glass window in a refrigerating mechanism that shows the amount of refrigerant or the oil in the system, or the pressure of gas bubbles in the liquid line.

signal circuit—Any electrical circuit supplying energy to an appliance that gives a recognizable signal.

silica gel—Chemical compound used as a drier.

silicon controlled rectifier (SCR)—Electronic semiconductor that contains silicon.

sill—Horizontal timber forming the lowest member of a wood frame house; lowest member of a window frame.

sine wave, ac—Wave form of single frequency alternating current; wave whose displacement is the sine of the angle proportional to time or distance.

single-phase motor—Electric motor that operates on single-phase alternating current.

skeleton construction—Buildings constructed of steel framing with the enclosure walls supported at each story.

sleeper—Wood strips embedded in concrete to provide a nailing base for the underflooring.

soffit—Underside of a stair, arch, or cornice.

soil stack—Vertical cast-iron pipe conveying sewage from branch pipes to house sewer.

solar heat—Heat from visible and invisible energy waves from the sun.

soldering—Joining two metals by adhering another metal to them at a low melting temperature.

soldier course—Course of brick consisting of brick set on end with the narrow side exposed.

solenoid—An electromagnet having a movable iron core.

soleplate—Horizontal bottom member of a wood stud partition.

solid bridging—Braces between floor joists consisting of short pieces with the same cross section as the joists.

spandrel—In steel skeleton construction, the outside wall from the top of a window to the sill of the window above.

specific heat—Ratio of the quantity of heat required to raise the temperature of a body 1 degree to that required to raise the temperature of an equal mass of water 1 degree.

split-phase motor—Motor with two stator windings. Winding in use while starting is disconnected by a centrifugal switch after the motor attains speed, then the motor operates on the other winding.

split system—Refrigeration or air conditioning installation that places the condensing unit outside or remote from the evaporator. It is also applicable to heat pump installations.

spot-face—The machining of a flat surface around the end of a hole to allow a bolt head or nut to seat squarely on the surface.

spray cooling—Method of refrigerating by spraying refrigerant inside the evaporator or by spraying refrigerated water.

spread footing—Footing designed for wider bearing on weak soils, often with reinforcing steel and of shallow depth in proportion to width.

squirrel cage—Fan that has blades parallel to the fan axis and moves air at right angles or perpendicular to the fan axis.

stack—Any vertical line of soil, waste, or vent piping.

standard conditions—Used as a basis for air conditioning calculations: temperature of 68 degrees Fahrenheit, pressure of 29.92 inches of mercury, and relative humidity of 30 percent.

starting relay—An electrical device that connects and/or disconnects the starting winding of an electric motor.

starting winding—Winding in an electric motor used only during the brief period when the motor is starting.

stator, motor—Stationary part of an electric motor.

steam—Water in vapor state.

steam heating—Heating system in which steam from a boiler is conducted to radiators in a space to be heated.

stirrup—Metal strap in a U-form supporting one end of a wood beam.

stratification of air—Condition in which there is little or no air movement in the room; air lies in temperature layers.

stretcher—Brick laid with its length parallel to the wall and side exposed.

strike—Door part of a door latch.

stringers—Members supporting the treads and risers of a stair.

strut—A compression member other than a column or pedestal.

studs—Vertically set skeleton members of a partition or wall to which the lath is nailed.

switch, ac general-use snap—A general-use snap switch suitable only for use on alternating current circuits and for controlling the following:

●Resistive and inductive loads (including electric discharge lamps) not exceeding the ampere rating at the voltage involved.

●Tungsten-filament lamp loads not exceeding the ampere rating of the switches at the rated voltage.

●Motor loads not exceeding 80 percent of the ampere rating of the switches at the rated voltage.

switch, ac-dc general-use snap—A type of general-use snap switch suitable for use on either direct or alternating-current circuits and for controlling the following:

●Resistive loads not exceeding the ampere rating at the voltage involved.

●Inductive loads not exceeding one-half the ampere rating at the voltage involved, except that switches having a marked horsepower rating are suitable for controlling motors not exceeding the horsepower rating of the switch at the voltage involved.

●Tungsten-filament lamp loads not exceeding the ampere rating at 125 volts, when marked with the letter T.

switch, general-use—A switch intended for use in general distribution and branch circuits. It is rated in amperes and is capable of interrupting its rated voltage.

switch, general-use snap—A type of general-use switch so constructed that it can be installed in flush device boxes or on outlet covers, or otherwise used in conjunction withwiring systems recognized by the National Electric Code.

switch, isolating—A switch intended for isolating a electrical circuit from the source of power. It has no interrupting rating and is intended to be operated only after the circuit has been opened by some other means.

switch, motor-circuit—A switch, rated in horsepower, capable of interrupting the maximum operating overload current of a motor having the same horsepower rating as the switch at the rated voltage.

switchboard—A large single panel, frame, or assembly of panels having switches, overcurrent, and other protective devices, buses, and usually instruments mounted on the face or back or both. Switchboards are generally accessible from the rear and from the front and are not intended to be installed in cabinets.

synchronous—Simultaneous in action and in time (in phase).

tachometer—An instrument for measuring revolutions per minute.

tap—A tool that cuts or machines threads in the side of a round hole.

tap drill—Drill used to form hole prior to placing threads in hole. The drill is the size of the root diameter of tap threads.

temperature—Degree of hotness or coldness as measured by a thermometer.

temperature humidity index—Actual temperature and humidity of a sample of air compared to air at standard conditions.

terneplate—Sheet iron coated with an alloy of four parts of lead to one part of tin.

terrazzo—Stone floor consisting of marble chips laid in concrete patterned with brass strips.

test light—Light provided with test leads that is used to test or probe electrical circuits to determine if they are alive.

therm—Quantity of heat equivalent to 100,000 Btu.

thermal cutout—An overcurrent protective device containing a heater element in addition to, and affecting, a renewable fusible member that opens the circuit. It is not designed to interrupt short-circuit currents.

thermally protected (as applied to motors)—When the words thermally protected appear on the nameplate of a motor or motor-compressor, it means that the motor is provided with a thermal protector.

thermal protector (as applied to motors)—A protective device that is assembled as an integral part of a motor or motor-compressor and that, when properly applied, protects the motor against dangerous overheating due to overload and failure to start.

thermal relay (hot wire relay)—Electrical control used to actuate a refrigeration system. This system uses a wire to convert electrical energy into heat energy.

thermistor—A temperature-sensitive resistor usually made from specially processed oxides of cobalt, magnesium, manganese, nickel, uranium, or mixtures of such substances. Thermistors can have either a positive or negative temperature coefficient of resistance.

thermocouple—Device that generates electricity using the principle that if two dissimilar metals are welded together and a junction is heated, a voltage will develop across open ends.

thermodisk defrost control—Electrical switch with bimetal disk that is controlled by electrical energy.

thermodynamics—Science that deals with the relationships between heat and mechanical energy and their interconversion.

thermometer—Device for measuring temperatures.

thermostat—Device responsive to ambient temperature conditions.

thermostatic expansion valve—A control valve operated by temperature and pressure within an evaporator coil, which controls the flow of refrigerant.

thermostatic valve—Valve controlled by thermostatic elements.

thermostatic water valve—Valve used to control flow of water through a system, actuated by temperature difference; used in units such as a water-cooled compressor or condenser.

three-phase—Operating by a combination of three alternating current circuits that differ in phase by one-third of a cycle.

thyratron—A gas-filled triode tube that is used in electronic control circuits.

timers—Mechanisms used to control on and off times of an electrical circuit.

timer-thermostat—Thermostat control that includes a clock mechanism. Unit automatically controls room temperature and changes it according to the time of day.

transducer—A device converting input into a different type of output.

transformer—A device used to transfer energy from one circuit to another. It is composed of two or more coils linked by magnetic lines of force.

transformer-rectifier—Combination transformer and rectifier in which input in ac may be varied and then rectified into dc.

transistor—An active semiconductor device capable of amplification, oscillation and switching action. The transistor has replaced the tube in most applications.

trap—U-shaped bend in drainpipe providing a space for water seal.

travertine—Porous marble building stone.

trimmer—Beam framing an opening in a wood joist floor supporting the header beam.

triode—A three-electrode electron tube containing an anode, a cathode, and a control electrode.

trusses—Framed structural pieces consisting of triangles in a single plane for supporting loads over spans.

ultraviolet—Invisible radiation waves with frequencies shorter than wavelengths of visible light and longer than X rays.

universal motor—Electric motor that will operate on both ac and dc.

urethane foam—Type of insulation that is foamed in between inner and outer walls.

utilization equipment—Equipment that uses electric energy for mechanical, chemical, heating, lighting, or other useful purposes.

vacuum—Reduction in pressure below atmospheric pressure.

vacuum pump—Special high efficiency compressor used for creating high vacuums for testing or drying purposes.

valve—Device used for controlling fluid flow.

valve, expansion—Type of refrigerant control that maintains a pressure difference between high side and low side pressure in a refrigerating mechanism. The valve operates by pressure in the low or suction side.

valve, solenoid—Valve actuated by magnetic action by means of an electrically energized coil.

vapor—Word usually used to denote vaporized refrigerant rather than gas.

vapor barrier—Thin plastic or metal foil sheet used in air conditioned structures to prevent water vapor from penetrating insulating material.

vapor lock—Condition where liquid is trapped in line because of a bend or improper installation that prevents the vapor from flowing.

vapor, saturated—A vapor condition that will result in condensation into liquid droplets as vapor temperature is reduced.

V block—V-shaped groove in a metal block used to hold a shaft.

velocimeter—Instrument used to measure air velocities using a direct reading airspeed indicating dial.

ventilation—Circulation of air, system or means of providing fresh air.

vermiculite—Lightweight inert material resulting from expansion of mica granules at high temperatures that is used as an aggregate in plaster.

viscosity—Term used to describe resistance of flow of fluids.

volt—The practical unit of voltage or electromotive force. One volt sends a current of 1 ampere through a resistance of 1 ohm.

voltage (of a circuit)—The greatest root-mean-square (effective difference of potential) between any two conductors of a circuit.

voltage to ground—In grounded circuits the voltage between the given conductor and that point or conductor of the circuit that is grounded; in ungrounded circuits, the greatest voltage between the given conductor and any other conductor of the circuit.

voltmeter—Instrument for measuring voltage action in an electrical circuit.

vortex tube—Mechanism for cooling or refrigerating that accomplishes a cooling effect by releasing compressed air through a specially designed opening. Air expands in a rapidly spiraling column of air that separates slow moving molecules (cool) from fast moving molecules (hot).

wall, fire—A dividing wall for the purpose of checking the spread of fire from one part of a building to another.

wall, retaining—Used to hold back a bank or solid mass of soil or water at the sides of the basement.

water-cooled condenser—Condensing unit that is cooled through the use of water.

water table—A projection of the wall at or near the grade line of a building. It is used to turn the water away from the foundation wall.

watertight—So constructed that moisture will not enter the enclosing case or housing.

watt—The practical unit of electrical power.

weatherproof—So constructed or protected that exposure to the weather will not interfere with successful operation.

web—Central portion of an I beam.

wet bulb—Device used in measurement of relative humidity.

wet cell battery—A battery having a liquid electrolyte.

window, bay—A window projecting outward from the face of the wall and built up from the ground.

window, oriel—A large bay window projected from a wall and supported by a bracket or corbel.

window unit—Commonly used when referring to air conditioners that are placed in a window.

wythe—Partition between flues of a chimney.

Index

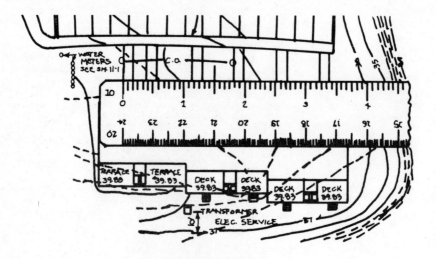

Index